BRIDGES

BRIDGES

their engineering and planning
including engineering basics, structures that keep
them up, hazards that threaten them, uses in
transportation, roles as American infrastructure,
costs and evaluation, environmental effects and
sustainability, and challenges of on-time delivery

George C. Lee
and Ernest Sternberg

Illustrated by
David C. Pierro

SUNY
PRESS

Published by
STATE UNIVERSITY OF NEW YORK PRESS, ALBANY

© 2015 State University of New York

For information, contact
State University of New York Press, Albany, NY
www.sunypress.edu

Production, Laurie D. Searl
Marketing, Anne M. Valentine

Library of Congress Cataloging-in-Publication Data

Lee, George C.
    Bridges : their engineering and planning / George C. Lee and Ernest
Sternberg ; Illustrated by David C. Pierro.
        pages cm
    Includes bibliographical references and index.
    ISBN 978-1-4384-5525-9 (hardcover : alk. paper)
    ISBN 978-1-4384-5526-6 (pbk. : alk. paper)
    ISBN 978-1-4384-5527-3 (ebook)
    1. Bridges—Design and construction. 2. Bridges—Planning. I. Sternberg,
Ernest, 1953– II. Title.
    TG300.L44 2015
    624.2—dc23   2014013135

10 9 8 7 6 5 4 3 2 1

In loving memory of Grace S. Lee,
to whom I owe all my accomplishments,
and who always cared about the education of students.

from George

To cousin Kati, of blessed memory,
who was killed in 1944 or 1945 when very young,
and could have become a builder of bridges.

from Ernie

CONTENTS

PART IV: CONCLUSION

TABLES AND FIGURES

TABLES

FIGURES

PREFACE AND ACKNOWLEDGMENTS

As we worked on our book, we consulted with Mr. Myint Lwin, director of the Office of Bridge Technology at the US Federal Highway Administration (FHWA). He told us of the two most serious challenges facing the highway system and bridges in particular. The first is the need for properly educated new professionals who can effectively design and manage the renewal of our aging system. The second is communication with the general public and with elected representatives, to make them aware that infrastructure investments require long-term commitment and the steady flow of resources.

We hope our book plays a part in answering both these challenges. We intend it to inspire students in search of satisfying careers to take up the study of bridge engineering and infrastructure planning. And we wish it to inform citizens and public officials about what their community will face when it decides whether to build or replace a bridge, and if it actually commits to doing so, the many complex tasks through which the project will be brought to completion. Oh yes, we are very glad to have as a reader anyone who is just curious. We are proud that over the course of our writing, and with assistance from the FHWA, our university has also established a master's degree program in bridge engineering, which is already graduating a new generation ready to face the future of aging infrastructure.

This writing project has received partial financial support from the FHWA (DTFH61-08-C-00012), the National Science Foundation PAES-MEN Individual Award (DUE0627385), the University at Buffalo Samuel P. Capen endowment fund, and MCEER, the multidisciplinary center for research on earthquakes and extreme events. To them we express our sincere gratitude.

The writing of an interdisciplinary book on bridges, by two authors with different backgrounds, one in structural engineering and one in urban planning, has depended on open dialog between us. With much discussion and with growing friendship we did indeed find the basis for mutually understanding complex topics well enough to put them into words we could each appreciate. We hope we have thereby been able to provide clear, well-rounded explanations to our readers.

Our ability to write this book has also depended on advice and assistance from friends, students, colleagues, and bridge-engineering professionals. Mr. Srikanth Akula did extensive analysis for us on American bridges (chapter 2), and later so did Mr. Sanket T. Dhatkar, who brought the analysis up to date. This effort was quite necessary because of the National Bridge Inventory's great complexity. We are obligated to Mr. Jerome O'Connor for his insistence that we redouble our efforts to make sure we had interpreted the inventory well, for his careful review of our chapter on bridge delivery, and for his suggestions for photographs.

Ms. Nasi Zhang helped us analyze and provide technically correct illustration of stresses and strains in a typical bridge under applied forces, helping bring chapters 5 and 6 to their present state, in which we strive for them to be accessible while remaining technically respectable. For chapter 9, to illustrate how planners analyze auto traffic for decisions about needs for a bridge, we hypothesized a simple place called Square City. The software with which we analyze Square City is known as "DynusT." We are grateful to Mr. Jinge Hu and Professor Qian Wang for developing the Square City model and running it for us. For the preparation of a bridge map appearing in chapter 8, we also thank Mr. Chao Huang and Ms. Paria Negahdarikia. Mr. Chao Huang also assisted us ably in organizing our many illustrations for publication.

For information on the bridge delivery process (chapter 10), we are indebted to Bruce V. Johnson, P.E., Oregon state bridge engineer, for his detailed knowledge, presented in an excellent slide presentation. For information on New York State highway development, we consulted the relevant environmental impact statements and received advice from Frank Billitier, P.E., and Norman Duennebacke, P.E. We are thankful to them for their help.

Early in our work, we consulted with Professor Alex Bitterman, who gave us valuable ideas for our work in general and for future illustrations. We were eventually joined by Mr. David C. Pierro, our illustrator, whose fine contributions are apparent throughout the book. Ms. Jane Stoyle Welch helped us greatly by insisting that we keep to a schedule and coordinate revision and illustrations. We want to conclude by expressing our thanks to University at Buffalo faculty members who gave of their time to advise us on the project. They include Brian Carter of the Department of Architecture; Himanshu Grover, Daniel Hess, and JiYoung Park of the Department of Urban and Regional Planning, and Niraj Verma, who has moved on from our university to head public policy studies at Virginia Commonwealth University. We would also like to express our appreciation to Myint Lwin and Phillip W. H. Yen for their detailed and useful manuscript comments. We now look forward to hearing from readers on whether they have been inspired to learn more about, and pay special attention to, bridges in the environment.

PART I

DECIDING ABOUT BRIDGES

ONE

CROSSING THE BRIDGE
BEFORE WE GET THERE

To have the kind of life we live in the United States and other advanced economies, where we enjoy freedom through mobility, which in turn fosters commerce, and where the built environment is safe for people, we depend on infrastructure. But it is a dependence about which we are largely unaware. Roads, water systems, ports, dams, electrical grids, and other physical public works function quietly in the background. They rarely attract attention because by and large they operate well.

Among the many systems in which we live from health care to finance, and among our daily worries from love to politics, public works provide some of our sturdiest and most reliable support. Love proves fleeting and papyri turn brittle, but Roman aqueducts still carry water and the US interstate system, like it or not, will dominate our landscape for a long time yet.

Disconcertingly to us, your authors, infrastructure may even seem boring. Streets and water pipes don't get to be national idols, don't have new upgrades released each year, can't be downloaded from your browser, and, when they're doing what they're supposed to, don't cause news. The infrastructure system's quiet dependability lets us forget what an enormous and complex technological achievement it is. Yet, on those who care to pay attention, it can exert a special fascination. In this book, we talk about one of these types of public works, the bridge. Why bridges?

The answer is in part personal: we like them, and one of us, George, has spent a large part of his career researching and teaching about bridges. More to the point, among types of infrastructure, bridges are the kind for which many people most easily acquire affection, and for good reason, though it is hard to express it. There is something stately about them.

3

Roads hug the earth's surface. Pipelines and tunnels burrow underneath it. But bridges soar through the air, without ever really leaving the ground. On beholders they make a distinct impression. Unlike buildings, which are more numerous but clad with outer surfaces that usually keep the underlying structure hidden, bridges reveal the structural principles that keep them aloft. They are the most visible expressions of engineering as art, or of architecture as science. Some are gateways to regions, symbols for entire cities, and world-renowned monuments in their own right. Some bridges, like some violin concertos, have magnificence that cannot be expressed in words.

While many kinds of human contrivances mar the natural landscape, bridges—even ones that are not particularly famous—are likely to complement it. They provide sequentially shifting panoramas for those crossing them, dramatic objects for those observing them from a shore or embankment, and framed horizons for those looking through or past them. Bridges as structural art are to be appreciated in their own right, but also as environmental art: pieces of artifice that enhance awareness not just of the artwork itself but also of the hills, chasms, torrents, skylines, or forests among which they are situated.

Before they can be art, they are economic infrastructure. They are essential because we move around on the earth and the earth's surface is, fortunately, not a flat and solid expanse. It has gullies, rivers, valleys, hills, swamps, crags, coves, and cliffs that must be crossed if we're to get about. Since we build roads and railways, it is often also wise to make them leap over each other instead of intersecting.

To accomplish that crossing by which it becomes an economic asset, the bridge must first be designed and built as a physical structure—which now needs definition.

WHAT IS A BRIDGE?

In movies when a galloping cavalry reaches a river, the riders inevitably coax the horses to swim across, just their heads above water, even if their mounts are in full armor. This way of crossing the river works, we suspect, only in the movies. Moses developed the method of getting the waters to part, a procedure that is no longer recommended since too many regulatory approvals would be needed. A ferry may be pleasant, if the waves are not too choppy and the wait at the dock not too long. In a pinch, and in the absence of a ferry or rowboat, a brisk swim might do; a catapult is best declined, even in desperation.

A bridge differs from the other ways of getting across in that it is a fixed structure that affords passage across; but, as a tunnel does the same by a rather different route, we have to add that the bridge reaches across by spanning a gap. By definition, then, a bridge is a structure that affords passage at a height across a gap. Let us now take the three pieces of the definition and consider them each, though in reverse order: the gap to be spanned, that which will make passage across it, and the structure that will support the passers' weight.

For the *gap* that the bridge crosses, a river most readily comes to mind, but it could just as well be a channel, lake, estuary, or the like. Or it may be a chasm, canyon, mining pit, ice crevice, or space between buildings. All these taken together still form a minority of the gaps that bridges cross. Many of the rest are the spaces between the raised sides of a roadway or railway. The curved ramp that raises or lowers traffic at highway interchanges is a bridge, too. So is the elevated highway, sometimes known as a viaduct, which spans the gap as it traverses a row of piers, sometimes casting its shadow over another highway running below.

That to which the bridge affords passage—well, it is people, vehicles, and the goods they carry, perhaps with livestock tagging along. Some bridges are solely for pedestrians and bicycles; a large number are for railways. In present-day America, that to which the bridge gives passage is overwhelmingly automobile traffic. Unless we specify otherwise, when we say "bridge" in this book, we mean one primarily meant to carry motorized road vehicles, though it may carry pedestrians and trains in addition.

The things that cross have weight and momentum. To afford them passage, the bridge must consist of an assembly of parts—a *structure*—that supports the forces acting on it. The structure must carry its own weight, stand up to the loads vehicles impart to it, and resist the forces of winds and waves and of the occasional errant barge that hits a pier. Those who would like to be informed about bridges should be able to understand the basics: the thinking by which engineers decide which kind of structure will safely carry the loads imposed on it.

THE BRIDGE DECISION

Even in a road transportation system as large as America's, we have far more bridges than most would guess, some 600,000 in fact. Every 500 or so Americans owns a bridge, or better put, each American owns a share in the nation's vast bridge portfolio. And that means many decisions have to be made about bridges, whether to build them, upgrade them, or close and replace them. At many places in America, every few years, citizens and their representatives, along with expert advisers, have to make such decisions.

We should pause, however, to consider whether it might be better to burrow underground to the other side than to span the gap above. It is rarely a good idea. Only in exceptional cases is a tunnel the right choice, for the very practical reason that tunnels are costly. Boring through rock and soil is expensive to start with; the price quickly spikes if the tunnelers run into geological formations they did not expect, something that readily happens underground, where no one is likely to have been before. Tunneling is dangerous for workers, further raising costs. Some danger persists even once the tunnel is in regular operation, not because the tunnel is likely to collapse, but because tunnel accidents are hard to clear, and tunnel fires and chemical spills are eminently to be avoided.

On the plus side for the tunnel, it may take up less space at the entrances than a bridge would, and that is a benefit in places where real estate is expensive. Tunnels are also preferred where storms make surface construction dangerous or where passing ships are so tall that the bridge would have to have very high clearance. Then again, if the channel to be crossed is deep, the tunnel must run correspondingly deeper, requiring long approaches (cars cannot handle angles of descent and ascent that are too steep), so that the tunnel may well have to be longer than a bridge would. At almost all places where there is demand to cross, the right structure by which to get across is the bridge, and in any case it is only bridges we study here.

Now, getting back to the bridge decision, here are the typical options. First, leave the old bridge alone, but increase maintenance, do some modest restoration, manage traffic better, and if possible persuade people to drive less. Second, reconstruct the bridge, by making structural improvements or expanding it. Third, if the bridge is too deficient, tear it down and replace it, though not in that order, since we need the old one to carry traffic until the replacement is finished. And fourth, the present bridges are fine, but demand has grown, so build a new one, adding to the region's collection of bridges. (If there is no present bridge, the choice is simpler, build or don't build.) Here are the choices once again: leave it and manage traffic, rehab or expand, demolish and replace, or build new.

Simple as the choices are to state, they are complex to make. They differ in important ways from other kinds of public policy decisions, though the differences are variations on a theme. All have to do with making early decisions.

Consider the annual town budget as a kind of public policy: if there is a shortfall, cut some programs or increase taxes. Skip to the local school district that's overenrolled: hire more teachers or maybe throw out some truants. Let's go to the bridge deemed dangerous from corrosion: now what? It takes years to build a new bridge. We have put this in a cavalier way, but the point is serious. When infrastructure has been poorly maintained for too long, or when traffic has built up too much, a patch-up here or there may work for a while, but the reckoning will come, and by then no quick fix will be possible. Good infrastructure decisions should be made before they are urgently due.

What is more, a bridge is a capital investment. To decide to build or reconstruct means that funds have to be expended this year for an item meant to endure and provide service over decades. We incur a large debt now, though we may not live long enough to experience the benefits. Unlike most policy decisions, which are driven by short-term calculation and the election cycle, infrastructure decisions (though they have current political costs and payoffs) have to be made for the long run.

As compared to other public concerns, like declining exports or increasing influenza cases, infrastructure is different again, because the problems it causes can be anticipated way ahead of time. Infrastructure causes problems not because we're surprised by the unexpected (there are exceptions, of course), but because we've been ignoring the expected.

Since it is expensive and very time consuming to fix the bridge when it is in danger of collapse, we should definitely not—in answer to this chapter's question—wait until we get to it to cross it. On questions of infrastructure planning, we should cross that bridge years before we urgently must.

THIS BOOK

The book that follows is a primer on the considerations at work when we decide whether to build or rebuild a bridge. Since many of the considerations resemble those for other kinds of infrastructure, some readers may also find in this book an introduction to infrastructure decisions in general, with bridges as the running example.

Throughout, we want to share our affection for bridges, which are among the most worthy and loved items in the built environment. The basics of bridge engineering are accessible to anyone who has spent a year or two in college, even if their major had nothing to do with science. To the viewer equipped with those basics, the bridge reveals much more than is otherwise obvious. Some may even become appreciators of bridges, hobbyists of sorts, stopping now and then to gaze at a fine structure. A few, we hope, will take up careers in engineering, planning, or architecture. (But we do not say much about bridge architecture because on that subject, as contrasted to bridge engineering and planning, there are already many books accessible to beginners.)

If we have done our work well, our book should also make clear that a bridge is a product of many professions and multiple analyses: bridge engineering for sure, but also financial analysis, transportation planning, environmental studies, and public policy making. Our book introduces many of the kinds of planning at work. For citizens concerned about making better bridges in their own communities, we offer our book as a guide.

Readers should be aware that, here and there, we give our views, a few of them controversial, on the directions in which we think bridge building and infrastructure policy should go. Where we express opinions (informed ones, we believe), the reader will be able to detect that from the way we write. Our most forceful claim is for the millennial bridge—but let us not reveal too much yet. We invite readers to find out what we mean.

We begin in the next chapter by counting America's bridges. We also estimate the number of sites, in a year, for which decisions have to be made about new construction or rehabilitation.

Then four chapters that follow should be read in a row: they are our engineering chapters. Chapter 3 provides the basics on the forces that bridge spans must resist to stay aloft. The next (chapter 4) explains how basic principles guide the engineer to design the types of bridges all of us observe on our travels. Though bridges are remarkably safe, their design cannot be based on certainty. Chapter 5 introduces the ways in which engineers manage to keep bridges strong, despite uncertainties. The most serious uncertainties arise from the possibility of extreme events, such as floods and earthquakes. These are the greatest challenges to bridge safety, and chapter 6 illustrates the ways in which engineers and other professionals strive to meet them.

Our series of chapters on bridge planning begins with the question: is the bridge worth building in the first place? Chapter 7 seeks to answer the question by introducing cost-benefit analysis for a bridge. This and subsequent chapters can be read in any order. The next (chapter 8) is on transportation planning and uses an extended example to analyze whether traffic pressures justify a new bridge.

The bridge to be built or rebuilt may well raise possibilities of environmental harm. Chapter 9 explains the process by which environmental impact is assessed and asks what could be meant by a "sustainable bridge." In chapter 10, our series on bridge planning ends by investigating a sometimes intractable problem: why a project often creeps along for a decade or more to get from initial studies to the day the ribbon is cut. We conclude the book with what we have already hinted about, our appeal for you to join us in advocating for bridges that span a millennium.

TWO

COUNTING OUR BRIDGES

In this chapter, we ask the question: just how often must big decisions be made about bridges? And to what extent is the United States facing a need for new bridges, bridge reconstruction, and bridge rehabilitation?

The place to go for answers is the National Bridge Inventory (NBI), a database maintained by the Federal Highway Administration to keep tabs on bridge conditions in the states. It assembles data each year from reports submitted by state transportation departments. As infrastructure is long-lasting, the national inventory changes fairly slowly, so the 2011 data, which we are using here, should remain a good indicator for years to come.

The fact that first strikes the eye is that there are over 600,000 bridges in the fifty states plus the District of Columbia and Puerto Rico. This is not even a full count, since the NBI counts only public bridges and leaves out tens of thousands of privately owned railroad bridges. Of the total in the NBI, 98 percent are road bridges, primarily meant to carry automobiles, trucks, buses, etc., though some also have lanes for pedestrians and tracks for trains or subways.

We classified the bridges according to length of the main span, so we could begin assessing the nation's bridge infrastructure challenge. We wanted to know, for example, how many are long enough that they could not have been built—and cannot be rebuilt—simply as girder (or beam) bridges.

To qualify for our classification, the span had to be greater than 20 feet, which is a short starting point since a span of that length barely crosses two road lanes. A 20- to 99-foot main span we classified as "short." If a bridge has a dozen spans, of which the single longest is 60 feet, then we still classified it as short-span even though the entire bridge is much longer. We classified a span of 100 to 329 feet as "medium," and 330 and over as "long." When a bridge exceeds 330 feet, it will almost always have to be designed as a truss, arch, suspension, or cable-stayed bridge. (We explain these types in chapter 4.)

Of the nation's bridges that fit our criteria, just under 87 percent have main spans in the short range (table 2.1). Even these modest structures make important statements in the landscape. In many towns in America, a 50-foot bridge can be a matter of pride, a public-expenditure concern, and a traffic choke point.

To be sure, longer bridges are the ones that garner the most attention. Of all American bridges, about 13 percent are medium-span, and one-fifth of one percent are long-span. Those numbers aren't peanuts. Medium- and long-span bridges taken together still amount to 61,000 structures, and many of them become deficient or obsolescent each year, raising the specter of rather expensive corrective maintenance or reconstruction.

The bridges aren't equally distributed around the United States. Of the states, Texas has the most, followed by Ohio, with Hawaii and Delaware at the bottom of the list. Alaska ranks low because of vast areas without roads. Cities are more likely to have higher densities of bridges because many sit alongside bodies of water, and almost all are highway and railway hubs, so they need overpasses and underpasses.

Of the top metropolitan areas (by population), the broad New York metropolitan area comes in second in its bridge endowment, with 7,952 bridges. Surprisingly, Dallas-Fort Worth comes in first with 8,888 bridges.

The St. Louis metro area has the greatest concentration of bridges per capita, with 163 per 100,000 people. Pittsburgh barely earns its billing as the "City of Bridges," coming out second with 158 bridges per 100,000 people (table 2.2). Sadly, the Los Angles metro area comes in quite low and may be said to be bridge-deprived. Bridge trivia this may be, but it also makes the point that some local governments face far more bridge decisions (relative to their population) than others.

Now we consider some of the basic reasons that people in an area might be confronted with bridge decisions.

Table 2.1. U.S. Bridges by Length of Main Span, 2011

Short: 20–99 ft	Medium: 100–329 ft	Long: 330 ft and longer	Total
397,494	60,016	1100	458,610*
86.7%	13.1%	0.2%	100%

Source: National Bridge Inventory (NBI).

*The NBI includes many bridges with main spans shorter than 20 feet. These we excluded from this table.

Table 2.2. Which metro areas have the most bridges? Ranked by bridges per 100,000 population, 2010

		# per 100,000 Pop.	Total bridges
1	St. Louis, MO-IL	163	4583
2	Pittsburgh, PA	158	3724
3	Cincinnati-Middletown, OH-KY-IN	146	3123
4	Dallas-Fort Worth-Arlington, TX	139	8888
5	Houston-Sugar Land-Baytown, TX	103	6145

Source: National Bridge Inventory

IS INFRASTRUCTURE AGING?

It requires little argument to win assent to the idea that the nation's infrastructure is aging, since everything is aging, including your present authors. For bridges, the pertinent question is whether they are on the average getting older—whether at some time the United States reduced its construction of new or replacement bridges, allowing older bridges to increase as a proportion of all bridges. If so, we have to be concerned about our aging bridges.

We tapped into the NBI to find out. Our findings tell a story that's more complicated than we expected. The number of bridges built shot up in the 1960s and has declined since then (figure 2.1). The declining number of

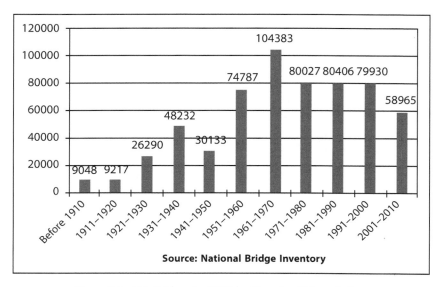

Figure 2.1. US Bridges in 2010 by Decade of Completion.

newly built bridges since the 1960s is not in itself a sign of neglect. Despite the decline in new completions, the bridge stock counted at (mostly) five-year intervals since 1992 (table 2.3) shows steady growth, with a small decline in the final half decade. The current stock of 605,086 represents over a three percent increase in just under twenty years. Some slowing in new bridge completions may be a good sign. It may well indicate that the nation's number of bridges simply has approached the saturation point—by the new century we had bridges at most of the places where we were ever likely to build.

So it's important to draw the right lesson here. The lesson is *not* that America has failed to build enough new bridges in the past three decades. Rather, it is that the spurt of bridge building in the 1960s and 1970s is coming due—these bridges are reaching an age at which they will pose ever more problems.

ARE BRIDGES DEFICIENT?

Old age is just a broad indicator that a bridge may require attention. Decisions on rehabilitation or replacement depend, of course, on actually observed problems. The NBI keeps track of problems, which it divides into two kinds, "structural deficiency" and "functional obsolescence."

Let's start with the former. For each bridge in the inventory, a state official fills out a form that evaluates the structural condition of the bridge components on a nine-point scale, starting with 9 for excellent. A score of 4 denotes deterioration, such as pieces falling off the structure. Skipping 3, we get to a 2, which indicates deterioration so severe that, subject to close monitoring, the bridge may have to be closed. With a score of 1 the bridge is in imminent danger of failing, so it should be closed to traffic, but may still be repairable. At the bottom, a 0 means the bridge is out of service and cannot be fixed. A bridge with a rating of 4 or below is labeled structurally deficient.

The bridge may, however, be obsolete even if it is structurally sound. For a particular type of road (say an interstate highway) and for a particular daily traffic load, engineers can consult national guidelines to decide

Table 2.3. Public Bridges in the United States, 1992–2011

	1992	1997	2002	2007	2011
US stock of bridges	585,830	596,632	604,233	612,205	605,086

Source: National Bridge Inventory

whether the lanes are wide enough; bridges having lanes that are too narrow by modern standards are considered obsolete. If clearances underneath for road traffic are too low by modern standards; if emergency road shoulders are insufficient or nonexistent; or if the approach roads to the bridge are subject to flooding or have curvature that is too sharp—for any of these reasons, too, a bridge is considered functionally obsolete.

So how do American bridges stack up? In making a judgment, we have to keep in mind that the data is collected by state agencies, which are required to use the same data when asking for federal highway funds. Following NBI instructions, a state official would have to list a bridge as structurally deficient even if the defect does not pose a danger of collapse, or list a bridge with narrow lanes as obsolete even if daily users consider it to be just fine. Then again, some of the deficiencies can be serious indeed.

The result is that 11 percent are structurally deficient and 13 percent are obsolete. Altogether 24 percent of the nation's bridges have one short-coming or the other or both (table 2.4). It's hard to know whether to read this result as good news or bad news.

The good news is that the percentage of deficient bridges has been declining (table 2.5). Structural deficiency has been dropping steadily from 20.7 percent of bridges in 1992 to 11.2 percent in 2011. Reasons may include increasing quality of the bridge stock brought about by new construction, and better maintenance and inspection. Over the same period, functional obsolescence has remained fairly steady, fluctuating at about 13 percent of bridges.

Despite improvements, 24 percent of bridges were still flawed in one way or another in 2011—that's almost 144,000 bridges! Now the bad news: the bridges built in the 1960s and 1970s are reaching an advanced age, suggesting an accelerating rate at which bridges will become deficient in the coming years (unless ever more is spent on keeping them in good repair).

IS TRAFFIC CONGESTION INCREASING?

A bridge may have to be upgraded or replaced, or an additional bridge may have to be built, for a reason other than deficiency: because it cannot serve the growing traffic pressure (i.e., it is functionally obsolete). Are

Table 2.4. Deficiency in Bridges, 2011

Not Deficient	Structurally Deficient	Functionally Obsolete	Total
461,197	67,526	76,363	605,086
76%	11%	13%	100%

Source: National Bridge Inventory

Table 2.5. Trends in Deficient Bridges

	1992	1997	2002	2007	2011
Number of bridges	572,196	582,751	591,220	599,766	605,086
Structurally Deficient	118,698	98,475	81,437	72,524	67,526
	20.7%	16.9%	13.8%	12.1%	11.2%
Functionally Obsolete	80,392	77,410	81,573	79,792	76,363
	14.0%	13.3%	13.8%	13.3%	12.6%
Not deficient	373,106	406,866	428,210	447,450	461,197
	65.3%	69.8%	72.4%	74.6%	76.2%

Source: National Bridge Inventory

bridges facing increased demands to carry traffic? Though we do not have reliable measurements of traffic exactly at bridges, we do know that through 2007 urban areas were indeed undergoing increased traffic congestion. That observation comes from the *Urban Mobility Report,* a study prepared by the Texas Transportation Institute and published in July 2009. Before accepting the result, the attentive reader must ask what "congestion" means, since it is by no means easy to define.

To gather their data, the Texas researchers studied conditions during peak travel hours, which they defined as 6 to 10 a.m. and 3 to 7 p.m. These are the hours during which about 50 percent of daily travel takes place—it is the time when the most demand is placed on road infrastructure. They then collected traffic data for these time periods at thousands of road segments in 439 urban areas.

For each lane in the road segments studied, they used computer programs to estimate travel times under free-flow conditions (no jams, breakdowns, crashes, or weather problems). With the collected traffic data, they then divided actual travel times during peak hours by the theoretical travel times under the free-flow conditions. The result was the "travel time index." If it were exactly "1," it would mean that traffic moved at the free-flow rate. But in all metro areas the index was higher than 1.

The Los Angeles metro area had the highest index—1.49—which meant that travelers on the average spent 49 percent more time traveling during peak hours than they would have under free-flow conditions. To exasperated Angelinos, the index may seem too low. But they must remember that the index includes travelers who hit the road at 6 a.m. and managed to escape the worst of the congestion.

The researchers then multiplied the average daily delay by the number of travel days per year to get average annual hours of delay per traveler. In the Los Angeles area it was 70, in Washington, DC, 62 hours, and in Buffalo, New York, 11 hours. In general, delay increased with size of metro area: the bigger the area, the more the delay. So the 14 very large metro areas averaged a delay of 35 hours per year, while the 16 small metro areas studied (from Charleston, South Carolina, to Wichita, Kansas) averaged 19.

Now we can get to our question: has congestion been increasing? As we see in figure 2.2, all sizes of metro areas have undergone increases in travel delays. In the 25 years after 1982, very large metro areas saw annual hours of traveler delay more than double.

It is a safe guess from this data that increased congestion overall means particular problems on bridges, because bridges are often traffic chokepoints (see chapter 8), where traffic congestion tends to be especially severe.

INFRASTRUCTURE CRISIS?

Overall, the United States since the 1990s has succeeded in reducing the percentage of structurally deficient bridges, and, of course, this is good news because structural deficiency implies dangers ahead. Then again, the spurt in bridge construction in the 1960s and 1970s is coming due. Many bridges are at an age at which they are accumulating expensive problems, which must be managed with corrective maintenance until reconstruction or replacement becomes essential.

That the percentage of obsolete bridges has fluctuated in the same range for these 20 years is less worrisome in itself. A minor shortfall in

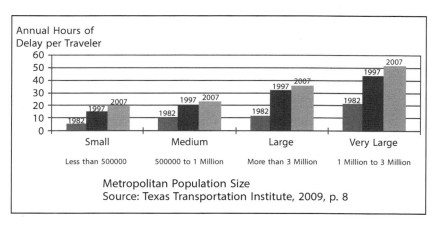

Figure 2.2. Trends in Travel by Metro Size.

achieving current standards may put the bridge in the obsolescent category while adding only marginally to the danger of travel. Then again, we have to keep in mind that the country's stock of bridges has grown. Even if the *percentage* of obsolescence remains steady, the *number* of such bridges has grown.

As we have seen, traffic is growing apace in cities and suburbs, especially in the largest metro areas. The demand does not necessarily have to be met with more bridges. Public transit, better traffic management, and incentives to get out of the car can reduce congestion while avoiding the expense of new structures. But we should not be too sanguine about possibilities for reducing car dependence. Energy crises and fuel-price spikes have come and gone, yet Americans have kept on driving.

Under the combined pressures of obsolete infrastructure and growing traffic demand, states and localities have continued to build new and rehabilitate old bridges. The NBI registers about 8,000 bridge completions per year in the United States, of which about 20 percent are rehabilitations and the rest are newly built or replaced, as shown in table 2.6. As we see in the table, rehabilitations have remained fairly level (with a peak in 2009), but new builds have been declining. With over 144,000 deficient bridges in America (of which 47 percent are structurally deficient and the rest obsolescent), we're chipping away at about 8,000 per year.

Additional bridges join the deficiency list each year, so we are always trying to catch up. And as the bridge stock from the 1960s comes due, the deficiency list will grow unless the United States accelerates the rate at which it builds new bridges. We are not in a bridge infrastructure crisis now, but it is around the corner.

Table 2.6. Bridge Building by Year

	New and Replaced	Rehabilitated	Total
2003	6641	2951	9592
2004	6504	1664	8168
2005	6130	1758	7888
2006	6182	1688	7870
2007	5334	1749	7083
2008	5364	1601	6965
2009	5368	2736	8104
2010*	4061	2421	6482

Source: National Bridge Inventory

*Incomplete data

THE BRIDGE DECISION

From the time that a bridge is proposed through final construction, the state or locality has to go through a labyrinthine process. When the bridge just uses an existing right-of-way and has no effects outside that narrow band, the process can take as little as three years. With lawsuits, budget shortfalls, and environmental controversies, the process can take two decades, if the bridge is ever built at all.

For the 5,000 or so new bridges for which construction is completed in a year (let's not consider rehabilitation now), easily another 20,000 to 30,000 are moving through the process from initial proposal, to community debate, to various stages of environmental study and construction.

What's more, at communities around the country, many more bridges pose problems of disrepair, deterioration, and traffic congestion. So there are additional tens of thousands of crossings over which debates, controversy, and budget battles swirl. What this tells us is that big bridge decisions are pretty common.

The decisions are made in large part by agency staffs and elected officials, but at various points in the decision process, citizens have important roles. For a citizen who wants to be an informed participant, basics come first. We need to know what goes into building a bridge that stands up against gravity's best efforts to pull it down.

PART II

BRIDGE ENGINEERING

THREE

UNDERSTANDING STRESSES AND STRAINS

═══════════════════════════════════

WHAT THE STRUCTURE MUST DO

Let us recall that the purpose of the bridge structure is to stand up to the forces that would drag it down. To engineers, these forces are *loads,* and the structure's capacity to withstand them is *resistance.* The engineer's fundamental job is to assure that loads imposed on the bridge do not exceed its capacity to withstand (to resist) the loads. The critical single lesson in bridge engineering, the indispensable idea, the one never to be forgotten, is that resistance should equal or exceed load. So a good place to start thinking about a future bridge is with an estimate of the loads it will have to carry.

To begin with, there is what is known as *dead load:* the structure's own mass along with those things permanently affixed to it. In almost all bridges, the greatest mass to be borne is that of the dead load. It may not be obvious that, for the vast majority of bridges, the paved deck on which one travels is actually not an integral part of the structure, but is rather carried on it—it is an additional item of dead load. Other dead loads are railings, traffic signs, traffic signals, and light poles.

Then there is *live load,* which is in turn divided into *stationary load* and *dynamic load,* the latter also known as *time-varying load.* The former consists of masses temporarily resting on the bridge—these might include cars and trucks stuck in a traffic jam or waiting at a tollbooth, heavy equipment (critical during construction and maintenance), people, vehicles' contents, and ice buildup.

Moving vehicles exemplify dynamic loads. When a car moves along the bridge deck, it bounces or vibrates, slightly jarring the structure each

21

time. When the car accelerates, it delivers a force on the bridge in a direction opposite the acceleration. Another way to put this is that the dynamic load changes in magnitude or intensity over the time that it is on the bridge. When the driver hits the brake, screeching to a halt, he imposes a more intense load than if he had slowed gradually. In normal traffic, these combined vehicular loads are asynchronous and intermittent, but on some occasions, as when a tractor-trailer jackknifes and the cars behind it simultaneously hit their brakes, the combined load jolts the bridge.

Another and extreme kind of dynamic force is a one-shot blow known as *impact load*. This may be an out-of-control truck that smashes into a column. It could also be a shake from an earthquake or a block of floating ice hitting a pier. Such matters are among the bridge engineer's greatest concerns, but we get to them in due course in chapter 6. In the meantime, let us just keep in mind the general lesson that the structure must bear the dead loads plus the various stationary and dynamic live loads to which it will be exposed.

STRESSES AND STRAINS

To begin even to assess the effects of a load on a structural member, engineers use a measure of what is defined as *stress*. Stress refers simply to load applied per unit area of the structural member. The reason for concern about stress should be clear. Imagine 100,000-pound weights placed on each of two upright cylinders of the same material. One has a diameter of 3.5 inches and the other of 7 inches. The 100,000-pound weight has much more of an effect on the thinner cylinder, but let us explain why. It is because a horizontal slice through the thinner one has an area of 9.6 square inches and through the thicker one of about 38 square inches. The thicker one has double the diameter, but four times as much cross-sectional area.

So as to not have to say "thousands of pounds," American structural engineers take recourse to a unit used by no one else: the kilo-pound, referring to 1000 pounds, and known for short as a *kip*. It is a hybrid between metric and the English customary system (to be contrasted to the traditional ton of 2000 pounds and the metric ton of 1000 kilos). In this parlance, an applied load of 1000 pounds per square inch is known 1 kip per square inch, and abbreviated 1ksi.

Recall that stress is defined as applied force per unit area. So the 100 kip applied to the thinner cylinder (about 10 square inches) exerts stress of 10 ksi, but on the thicker cylinder (let us round it off to 40 square inches), about 2.5 ksi. The stress from the load is four times greater on the thinner than the thicker cylinder (figure 3.1).

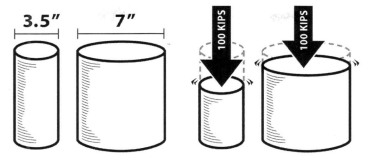

Figure 3.1. A 100-kip load imposes more stress (causing strain) on the thinner cylinder.

Engineers compare the stresses imposed to the *strain* undergone by the structural components. In our example, the component is put under a stress of 100 kips. The resulting deformation that the column undergoes is *strain*, which is shortening per unit length. Strain is measured by the number of inches the object deforms divided by its original dimension. If that is too abstract, wait a moment longer for the next section.

But first let us prepare you for potentially difficult terminology. Let us say that we impart a load on top of a column—this is called a "compressive force." The effect of the force depends on many factors, including the column's thickness. When measured by unit area to which the force is imparted, the same load can be said to impart a "compressive stress." And the shortening that the column undergoes per unit of original height is the "compressive strain."

These concepts are so closely related to each other that engineers sometimes loosely use them interchangeably. For short, remember that a load stresses a component and the component strains under it.

Stresses and resulting strains in a structural member come in several types, depending on the kinds of forces applied to it. In fact, they come in five *basic* types, plus combined types. The one we have discussed so far is compression; the others are tension, shear, bending, and torsion.

COMPRESSIVE FORCE: PUSHING ON A COLUMN

Let us first examine the effect of *compressive force* acting on a slender upright column. Do not worry yet about the material from which the column is made, except to say it is structural material, meaning that it exhibits strength in response to loads placed on it. For now, we assume that the material behaves homogenously throughout and that the column rests fixed on an imaginary platform that will never allow it to sink downward.

Since columns are frequently used to support bridge loads, we should be rather interested in how the column behaves when a load is acting on it. We now put our 100,000-pound (100-kip) weight on top of a column, whose cross-sectional area (the area of a horizontal slice through it) is 10 square inches. We put it there in a perfectly gentle way so that we need consider only the pushing effect of the static mass itself, and not the dynamic effects that would occur if we were to drop it into place. Once there, the load exerts a compressive stress (downward) on the column (figure 3.2). Once again, the load exerts a per-unit stress of 10 ksi.

What happens to the column? Since it is fixed onto its platform, it cannot move downward. No material is perfectly rigid, so the column undergoes a deformation: as we expect, it becomes shorter. We could imagine that the fibers in the material are getting pressed together—but that is just a convenient mental picture; we should not think that molecules actually behave that way. The shortening per unit length is the *strain* and is measured as a ratio between the amount of shortening and the original vertical length.

As progressively heavier loads are placed on top of the column, we can expect ever more shortening, until such the load reaches a threshold, this being the highest stress the column can bear. This is known as the material's "ultimate strength," beyond which danger lies. If the column is short, the

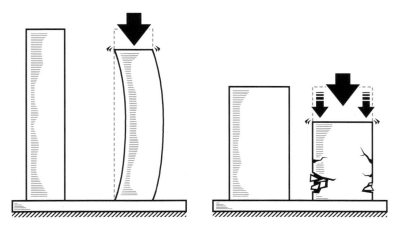

Figure 3.2. Beyond a column's ultimate strength, the load causes buckling in the tall column, but crushing in the short column.

compressive force is likely to act uniformly throughout the cross section of the column, crushing the material. For a tall column, the compressive force may cause the column to buckle even before its ultimate stress is reached.

If we compare columns of the same cross-sectional area and same material, each subjected to the same compressive force, the taller the column, the more likely it is to buckle. Since the engineer's primary job is to ensure that the column's resistance exceeds its load, she must anticipate ultimate strength ahead of time, and ensure that loads above this critical value are prohibited. If the load cannot be reduced, she may select a column that is shorter, has a larger cross-sectional area, or is made of stronger material.

TENSILE FORCE: PULLING ON A CABLE

We now turn to tension or *tensile force*, which acts by pulling on a material. Let us consider a load, once again 100 kips, suspended from a steel cable (figure 3.3). This is a worthwhile subject for bridge builders because

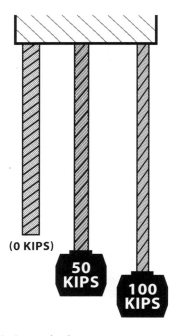

Figure 3.3. Larger loads impose greater tensile strain.

suspension bridges support loads through cables called hangers, composed of many entwined wires, extending vertically downward from a main cable, which is made up of even more wires. In our imagination, we make sure that the hanging cable is so firmly fixed to its main cable that it cannot come undone. The downward stress of the load is measured once again in pounds per square inch of the hanging cable's cross-sectional area (that's a horizontal slice through it). If the area is 5 square inches, then the 100-kip load imposes a *stress* of 20 ksi. The stress has the effect of deforming the cable: stretching it downward. The *strain* is the measure of that stretching: it is the ratio of the cable's increase in length to its original length.

As we increase the weight of the suspended object, stress increases, correspondingly increasing the strain. It is as if the fibers inside the material were being stretched further apart—but again we should not think that somehow we are visualizing actual molecules. Past a maximal value, the steel cable suddenly elongates precipitously. If the room under it is great enough, it elongates until it snaps, hurtling the suspended load to the ground. That of course is the outcome bridge designers must avoid. The designer must restrict the load to start with, or otherwise increase the cable's diameter, add additional wires, or use stronger wire.

Discovered by the English scientist Thomas Young (1773–1829), "Young's modulus" (the word *modulus* just means "little measure") demonstrates that each additional increment of stress causes a proportional increment of strain. More pulling load on the cable causes more strain on the cable—up to the point at which it lengthens precipitously per unit of added stress. The stress threshold that causes this change in the cable's behavior is known as the "yield stress."

Likewise, more compressive force on a column causes proportional increments of strain, until the yield threshold at which the column begins to buckle or get crushed. When planning for the loads to which a cable or column can be safely subjected, the engineer refers to information on the yield stress at which the material no longer responds proportionately to added force (figure 3.4).

Note that under tensile pulling, there cannot be an outward bulging of the material—there cannot be buckling. If we compare steel bars, one under compression and one under tension, both having the same cross-sectional area, the bar can undergo more tensile stress before snapping than compression before buckling.

This may be the point at which to add that stiffness is not at all the only desirable quality the designer looks for. Whether it is meant to resist compression or resist tension, the structural member must not retain deformation (shortening in one case, lengthening in the other) after the live load has been removed. The ability to bounce back from deformation

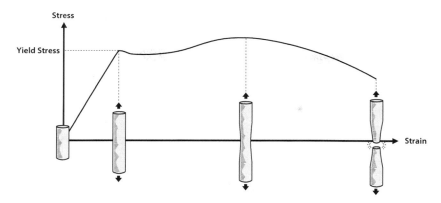

Figure 3.4. From original size (0), the cable stretches proportionately to applied stress, until yield stress (A). Beyond that threshold, the cable deforms permanently (B), and eventually snaps (C).

is known as elastic response. Upon having crossed a suspension bridge and thereby having slightly lengthened the wires and cables on which the deck rests, you want them to spring back to their original form. An engineer's way to say this is that the material must stay within its elastic limit. If not, the next person crossing the bridge will elongate the cables even more, causing permanent deformation, which can become dangerous.

SHEAR FORCE: SCISSOR EFFECT ON A BEAM

In our examination so far, loads have been applied in a direction directly in line with the structural member's longitudinal axis. The compressive force has pushed downwards directly in line with the column's axis; the tensile force has pulled down directly in line with the cable from which it was suspended. The stress we investigate now—*shear force*—is applied transversely: at a 90-degree angle to the structural member it is stressing.

The structural member we have in mind now is a slender beam, rectangular in its cross section, laid horizontally over a gap, each end firmly attached. Assume that you are standing in the gap, staring at the side of the beam. Let us name the beam's three dimensions: the *length* is the longest dimension, extending horizontally in front of you. *Depth* is the up-down distance. And *width* is the distance in the directions to and from you (figure 3.5). To make it resemble a simple beam bridge, we make the beam's depth greater than its width. This reflects the discovery that wood bridge builders made long ago, that if the wooden planks were laid flat (the width of each

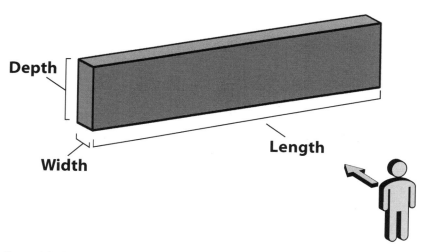

Depth

Length

Width

Figure 3.5. Terms for a beam's three dimensions, from an observer's point of view.

plank greater than its depth), they would flap up and down with the passing load. But if enough planks were available and were laid next to each other while resting on their narrow edges (width now much narrower than depth), the structure would become much stiffer—it could resist far greater loads.

A stiff beam exhibits a kind of stress behavior that most of us would not think of. Called *shear*, it occurs when forces push in opposite directions: not at directly opposed points on the same cross-sectional plane (that would just be compression) but at opposite points on adjacent planes (figure 3.6). For example, consider the effects of a pair of pliers on a sheet of metal as compared to a pair of shears. As you press together with the pliers on the sheet, forces converge from opposite directions on the same area in the metal, causing ordinary compressive forces. By contrast, if you press with the shears on the sheet, the area the downward force presses is separated by a very small space from the area the upward force presses. The sheet of metal will be cut—or sheared—apart.

But even a beam on which no scissors act, one that bears only its embodied dead load, undergoes shear strain. This is important, so we should try to picture it. Imagine that the beam is an assemblage of cards glued together and placed horizontally across two bricks, directly in front of us, left to right. As we follow the cards left to right, we see some cards that are resting directly on the left brick, until we reach the first card suspended over the gap. As compared to the previous card over the brick, the first card over the gap will tend to slip downward. An analogous process occurs in a beam laid across columns above a river: the beam's fibers adjacent to the column undergo shear strain (figure 3.6).

The shear strain occurs even if we have in mind just the dead load. If we put a live load (a paper weight) on the card bridge just past the left brick, the shear strain is even greater.

As the applied load increases, we can observe an equivalent to Young's modulus, but now for shear. As the load increases, the shear strain at first reacts proportionately, slipping in direct proportion to the weight placed on top of it. Beyond a threshold however, the slippage becomes excessive, endangering the bridge.

Our pack of cards gives a sense of the stress undergone by a beam extending from one riverbank across a column in the middle of the river to the other bank. The beam experiences high shear strains at fibers extending just past the left bank, just past each edge of the column, and just before the right bank.

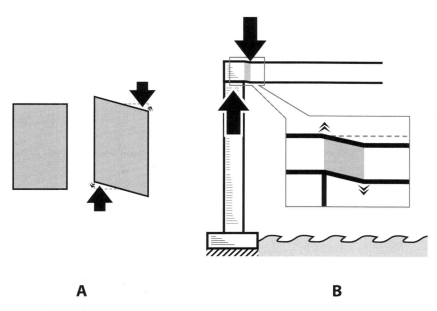

A   **B**

Figure 3.6. (A) Shear forces applied to a component. (B) Shear strain experienced by a beam at its juncture with a column.

BASIC BENDING

Say we are holding both ends of a bar, trying to bend it with equal effort by both hands. The result is a form of bending, but it is best to think of it as "basic" bending, to differentiate it from more complex kinds of bending.

To have an intuitive understanding of what happens, it is no longer adequate for us to think of the bar as a pack of cards. So let us instead imagine the bar as composed of very many equal-sized tiny cubes, connected by imaginary glue. Let us also say that no amount of applied force can change the cubes' shape. Rather, forces applied to the bar cause the cubes to change their relative locations (push them closer together or pull them farther apart), deforming the bar.

Let us start with the original bar, before it is bent. The imaginary glued space between any two adjacent cubes is equal throughout the bar. As we grip each end and press in a way that rotates each end downward and inward at the same time, the middle of the bar bends upward. Let us now examine what is happening. The cubes on the upper surface of the bar get stretched apart (the glue between cubes is distended). The cubes on the lower surface are squeezed closer together. Another way of putting this is that the upper layer of cubes is in tension and the lower layer in compression (figure 3.7).

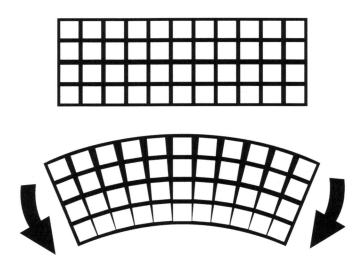

Figure 3.7. Under bending, the imaginary cubes at the beam's upper surface are stretched apart (undergo tension), and at the lower surface get pushed together (are compressed).

Assume now a vertical cross section through the bar and let us move along this cross section, starting on the upper surface. At the very top layer, the cubes are being pulled apart the most (are most under tension). As we move down a few layers, the tension declines. In the middle layer of cubes running along the bar, there is neither tension nor compression. As we proceed to the lower layers, the cubes start to get squeezed together (are under compression). The greatest compression is in the bent bar's bottom layer. As we bend the structural material ever further, we again run to limits beyond which further bending causes damage.

To understand why we call this "basic bending," imagine that all we have said is occurring in outer space, with no effect of gravity at all. Here on earth, when we lay a long beam over a crevice, the actual bending we observe is rather more complex than this basic bending we have just described. In fact, actual bending is likely to combine several of the kinds of forces we have described here.

BASIC TORSION

Now to our fifth and final kind of application of force to a structural member, consider a solid, circular cylinder, rigidly attached at one end to a wall, a wall that won't budge. In a process known as *torsion*, we now twist the cylinder's free end (figure 3.8). To understand what happens, let us again think of the cylinder (remember, it is solid, not hollow) as being composed of tiny cubes all connected to each other with imaginary glue. These cubes are so tiny that the cylinder's surface looks smooth; if a very thin cross-sectional slice is taken out of it, it is a circular disk composed of one layer of very tiny cubes.

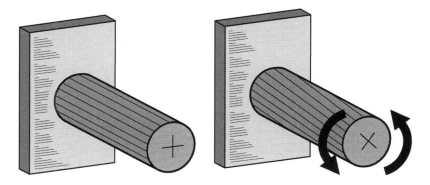

Figure 3.8. Cylinder undergoing torsion at its free end.

From the wall stretching out toward the free end, the cylinder is, by this abstract model, composed of successive layers of disks, each disk composed of many cubes. As we twist at the free end, the disk closest to us turns the most. Each successive disk-layer turns somewhat less, until the very final disk, the one closest to the wall, which hardly turns at all. Somewhat ambiguously, the term *torsion* has two meanings. It means either the torsional force applied, or the torsional strain that the cylinder undergoes, or both, depending on context.

Note that there is a correspondence between shear strain and torsional strain. In the former, as force is applied transversely to a structural member, layers of material slide against each other. In the latter, one disk layer rotates against the adjacent disk layer. As one disk rotates more than the disk adjacent to it, the glue between them is distended. This is the reason that torsion can be said to consist of many disk-to-disk shear effects added up along the cylinder's length.

Up to now, we have very briefly described the kinds of stress (and resulting strains) produced by five different forces: compressive, tensile, shear, basic bending, and basic torsion. However, the strains that real-life structural members—real beams or columns—undergo reveal the effects of multiple forces in combination.

COMBINED STRESSES IN ACTUAL BENDING

Let us again place a load on a beam horizontally suspended on two supports, but this time let us put the load in the middle. As one would predict, the load makes the beam bend downward. We would expect that intuitively from our ordinary knowledge of the physical world. What is not obvious is that the bending is not basic bending but a more complex process.

Under basic bending the beam's top layer of cubes is pressed together. This upper perimeter layer actually gets shorter. The cubes in the beam's lower layer are stretched apart; the lower perimeter layer becomes longer. The beam has undergone both compressive stress and tensile stress (figure 3.9).

To make the picture more complicated, this beam also undergoes shear strain. To picture it, let's remember that our beam is composed of tiny cubes arranged such that they can be visualized in either vertical or horizontal planes. Now we put a load on the middle of the beam's upper surface. The vertical layers of cubes directly under the load now dip downward, slipping past the adjacent vertical layers (the ones just beyond the load). The layers under the load are shearing against the adjacent layers.

Under bending force, therefore, the beam undergoes three kinds of strain. First, at its upper surface, it experiences compressive strain, which can

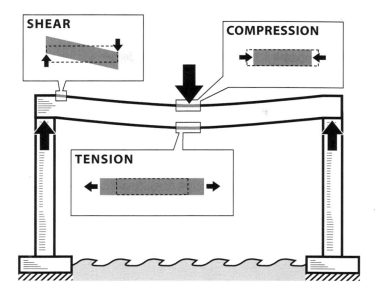

Figure 3.9. Actual bending on a beam bridge includes normal bending (compression on the upper surface, tension on the lower) and shear.

(beyond some limit in material strength) lead to the puckering up (eventual crushing) of the material. Second, at its lower surface, it undergoes tensile strain, which can (again, past a limit) cause cracks until the material finally snaps. (Both compressive and tensile strain are referred to by engineers as "normal strain" because the shortening or lengthening happen in the beam's longitudinal direction.) Third, cross-sectional plates under the edge of the imparted load slip past each other until a critical threshold, past which the beam undergoes *shear failure*.

With a beam of given cross section and given material, the amount of compressive, tensile, and shear stress that a static load imparts depends on the load's weight, the length of the beam, and the load's location along the beam. The bridge designer's job is to ensure that the stresses developed in a structural member do not exceed the material's several thresholds.

BENDING MOMENT:
THE RELATIONSHIP BETWEEN DISTANCE AND APPLIED FORCE

As we have put it so far, the extent of bending in a beam depends on the magnitude of the load applied to it. Now we must add that the extent of bending also depends on the length of the span between the beam's two supports.

It seems obvious that the longer the beam, the more it will bend under the same imparted load. But why? For the same reason that, when you hold a stick in two hands, firmly gripping the stick at each of its ends, it is easier to bend a longer stick. (We're assuming that the stick's diameter remains the same.) The stick's length gives leverage by which to bend it, just as a lever imparts leverage to someone trying to tilt a heavy object. Called "moment" by engineers, this is a force imparted in such a way as to cause rotation about an axis.

To understand, imagine that the beam that was originally fixed at each end to a wall is sliced apart at the middle, dividing it into two cantilevers. Each cantilever is still firmly attached to its wall. A live load (a weight) applied on top of the far end of either cantilever will tend to rotate the beam downward. For a given applied load, the longer the cantilever, the greater the tendency to rotate downward. For a given length, the greater the applied load (the weight), the greater the tendency to rotate downward. In either case, we observe downward bending, with the displacement (from horizontal) increasing toward the free end.

This downward bending combines several kinds of strain we have already discussed. First, as we move from the wall toward the free end, each vertical layer of cubes has slid slightly downward as compared to the neighbor closer to the wall. In short, each has undergone *shear strain*. Second, the top layer of cubes has been elongated, causing tensile strain; it is greatest right next to the wall. Third, the bottom layer of cubes has shortened, causing compressive strain; it, too, is greatest right next to the wall. The extent of the dip from the horizontal (we have called this the *displacement*) at the free end shows the accumulated effect of all these strains (figure 3.10).

The critical payoff for the structural designer is obvious: the longer the cantilever (everything else being equal), the greater the downward dip, and the greater the strain (compressive and tensile) next to the wall. Engineers explain this phenomenon as *moment*: the relationship between distance and the applied force exerted by the load at the free end. A short cantilever with a large applied load may have moment equal to a long cantilever with a small applied load.

Let us reconnect the two cantilevers, making them into a beam-bridge again. Now the beam is sturdier in the middle because the former cantilevers are joined. However, each half (each former cantilever) still experiences effects of moment. The longer the beam (other things being equal), the greater is each former cantilever arm's moment. The greater the moment, the greater the bending, and hence the greater the various strains (tensile, compressive, shear) occurring in the beam.

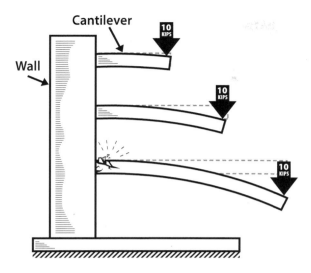

Figure 3.10. Effects of bending moment, shown by live loads of equal weight applied to ever longer cantilevers.

COMBINED TORSION AND BENDING

When we examined actual bending, we imagined a load applied directly on top of the beam's upper surface and (though we did not specify earlier) evenly spread across the beam's width. Recall that the beam is fixed to supports at each end. To take our discussion a step further, we need a clearer vocabulary by which to explain the location of the load on the beam's upper surface.

Imagine that you are in the middle of the gap the beam is spanning, looking directly at the side (the dimension we measured as *depth*) of the beam at eye level. If the upper face of the beam were to tip toward you, we can say it would tip "forward." We once again apply the load at the middle of the beam's longitudinal upper face. But, now we position the load at the forward edge of the upper face, the edge nearest to you. We assume that the ends of the beam are properly held in their original position. The load now causes downward bending as before, but it also causes torsion.

The load causes downward bending, with the accompanying (basic) bending strain, compressive strain, tensile strain, and shear strain. It also causes the forward upper edge to tip forward (toward us). It tips more in the middle of the span than at the edges. To understand, let's imagine, one more time, that our beam is composed of thousands of vertical layers of cubes. As the load bears down on the beam's upper forward edge, it tips the vertical layers under it toward us, and tips the adjacent layers slightly less, and the layers next to them even less. Now, in addition to causing basic bending, compression, tension, and shear, the load imparts a kind of twisting.

Note that when the middle layers angle forward more than their neighboring layers do, they slip past their neighbors, stretching the imaginary glue between them. They undergo torsion. With enough rotation, torsional strain can combine with several other kinds of strain to cause failure in the structural member.

TOWARD ENGINEERING DESIGN

When an engineer designs a particular bridge type, she must be aware that the future structure will encounter multiple loads: the structure itself, the deck and pavement, automobile traffic, intermittent heavy truck traffic, wind, earthquake, loads imposed by thermal expansion, and effects of settling soil, among others. These loads will exert varied kinds of stress at each location along the bridge. The engineer's task is to ensure that the structure transfers the various loads to the ground, without failure or collapse.

Toward that end, each structural member must be selected so as to withstand the compressive, tensile, shear, basic bending, torsional and combined forces that the loads will exert on it. What is more, not just each component on its own, but the whole interconnected assemblage that makes up the bridge structure must stand up to the loads it will carry. To turn these abstract concepts into real structures, engineers have relied on series of typical structural types. It is to these that we turn in the next chapter.

Sources and Further Reading

Mario Salvadori's *Why Buildings Stand Up* (New York: W. W. Norton, 1980) is a well-known and pleasant introduction to structures underlying architectural form, though a few of the book's parts (as about earthquakes) are no longer considered correct. A step up in complexity but still highly accessible is Waclaw Zalewski and Edward Allen's *Shaping Structures: Statics* (New York: John Wiley, 1998). A more challenging book but one still accessible with elementary mathematics is R. E. Shaeffer, *Elementary Structures for Architects and Builders*, 5th ed. (Upper Saddle River, NJ: Pearson/Prentice Hall, 2007).

FOUR

BRIDGE TYPES AND SITES

Any bridge is an assemblage of structural members, such as beams, piers, slabs, cables, stays, and arches. The bridge designer must select the members' materials, sizes, and shapes and then properly interconnect the members, so as to provide safe but cost-effective crossings. Through thousands of years of human experience with bridges, designers have discovered and sometimes rediscovered that certain structural configurations are especially well suited to the task. These are the bridge types to which we now turn.

STRUCTURAL MEMBERS: MATERIAL AND SHAPES

Let us begin by saying a few word about the material from which structural elements, such as beams and columns, are made. The material should be *stiff*: it should not deform too much under stress. Having undergone stress, the material should also be *elastic*: once the live load is removed, it should bounce or twist back to its original shape.

Past its limit of elasticity, it should not snap or break suddenly, as a rubber band does. Rather it should have the additional feature known as *ductility*, by which stress beyond a given limit permanently distorts the structural member. Under excess bending forces, for example, we want the material to retain its distended shape, to serve as a clear indication that it has undergone excess strain.

By such standards, structural steel performs very well. A properly manufactured steel cable when pulled (tension) can withstand well over 50 kips (50,000-pound) static load per square inch. Recall the hypothetical cable mentioned in the previous chapter as part of our discussion of tensile stress. It has a cross-section of 5 square inches and is subjected to a 100-kip load. Hence, the cable is subjected to 20 kips per square inch, an amount that is well within the capability of most steel cable. Steel is also well-known for its elastic behavior; it quickly regains its shape after stress. When excessively stressed, steel also exhibits ductility, reducing the chance of catastrophic failures. As fine a structural material as steel is, it has the disadvantages that it is expensive and undergoes corrosion over time. To retain its excellent qualities, steel requires corrosion protection and preventive maintenance.

Made from sand, crushed stone, pebbles, and a slurry mixed from cement and water, concrete is less costly than steel, is made from natural materials found all over the earth, is easily cast into needed shapes, and has excellent resistance to compressive forces. Arch bridges that depend mainly on compressive forces are readily adapted to construction from stone or concrete, and have withstood the test of time. However, concrete's capacity to withstand tensile stress is much lower than its resistance to compression.

Reinforced concrete remedies this problem. Consider a concrete beam to be used to withstand downward bending force. Under such force, the beams' bottom will undergo tension, for which it is not well suited. However, a steel reinforcing bar encased in the bottom of the concrete will add tensile capacity inherent to steel, while the rest of the concrete retains its normal capacity to resist compression. In most cases, the reinforced concrete beam provides the needed resistive property at lesser cost than steel (figure 4.1).

While engineers must choose the structural elements that most safely resist load, they are also obligated to do so at least cost. If the structure is to be economical, they should select structural members that are as slender—use as little material—as possible. It helps the purpose that the solid beam is wasteful at carrying loads. Economies are attained by scooping out some of the beam's internal contents.

To understand, let's return to the horizontal beam undergoing bending stress, and observe what is happening at its cross section. As it bends downward, we observe that the compression forces are taken up by the top layers in the beam and the tensile forces by the bottom layers. The middle

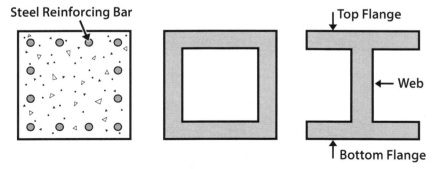

Figure 4.1. Cross sections of reinforced concrete beam (left), steel box beam (middle), and steel I-beam (right).

layers carry far less stress. This is a well-understood phenomenon. By carving out the beam's middle layers, we can reduce the beam's weight while still retaining a high proportion of its bending capacity.

One of the efficient shapes for doing so is the I-beam, composed of a top flange for handling compression, bottom flange for tension, and what is known as the *web* for connecting the upper and lower flanges and to take up the shear stresses (figure 4.1). It is worth taking a moment to note this specialized use of the word *web*, which has nothing to do with our usual image of it. An alternative to the I-beam is the box beam, which has the cross-sectional shape of, well, a box. The box beam of course has the upper and lower flanges, plus a web on one side and another web on the other side (figure 4.1). Its advantage is that it has better resistance to torsion forces than the I-beam does. Its disadvantage is that it is more costly to fabricate.

Though the interesting bridges one sees are composed of wires, cables, beams, and columns, we begin our discussion of bridge types with the most modest kinds of bridges, the culvert and slab.

CULVERT AND SLAB BRIDGES

Though rarely discussed, culvert and slab bridges are ubiquitous on American roads. The culvert may be a concrete or rock structure with earth covering, above which a road runs. Or it may just be a large pipe crossing under the road. Either way, its purpose is to allow surface water to drain, maintain a channel for a stream, give room for town utilities, or allow animals to migrate. Some masonry culverts are distinctive architectural remnants, ignored by all except a few—very few—culvert aficionados. If it requires a special taste to appreciate a culvert, all the more so a slab, which to our knowledge has no fans (figure 4.2). A slab bridge is just what one would

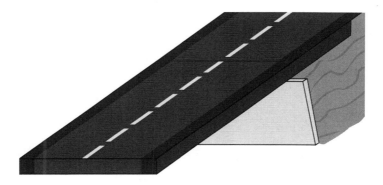

Figure 4.2. Slab bridge.

guess, a flat block of material laid across a minor span, perhaps across a two-lane road. For short spans of 20 feet or so slabs and culverts are the most numerous types of bridges in America.

From a bridge designer's point of view, slabs and culverts are not exciting affairs, because a short span limits the stress that loads put on the structure. Prefabricated parts easily fit the need. That said, road managers are forewarned not to neglect them. Floods have been known to wash culverts away, taking chunks of road with them along with the lives of unfortunate motorists, and slabs do crack and collapse after one-too-many a truck has bumped over them. As subjects of engineering they are unexciting, but as objects for attention by maintenance crews they are important indeed.

GIRDER BRIDGES

At a time lost to eons past, a human or hominid ancestor constructed the first bridge by laying a log across a crevice. It was a girder bridge or, equally, a beam bridge: a longitudinally slender material resting on supports (the ground on both sides of the crevice) on to which it directly transmitted the suspended live-load force, that being the weight of our ancestral engineer. Modern girders function on the same principle, but the supports on which they rest are likely to be piers, which act as force-transmitters to the ground. That's also the principle holding up multiple-span bridges (viaducts), which consist of girder spans laid across linearly arranged sets of piers (figure 4.3).

American highway bridges commonly rest on I-girders or box-girders several feet deep (figure 4.4). What is the difference between a girder and a beam? A girder is a kind of beam. Many modern bridges are constructed of beams of several sizes, sometimes placed perpendicularly to each other. Among the beams, the girders are the main load-carrying members and are aligned in the direction of the traffic. Where two girders are the main load-carriers, other beams may be placed perpendicularly between them to keep them rigid widthwise.

In suspension bridges and cable-stayed bridges, the deck is suspended below the superstructure and pulls down on it. In a traditional masonry arch bridge, the deck rests on a substructure and pushes down on it. In the

Figure 4.3. Viaduct.

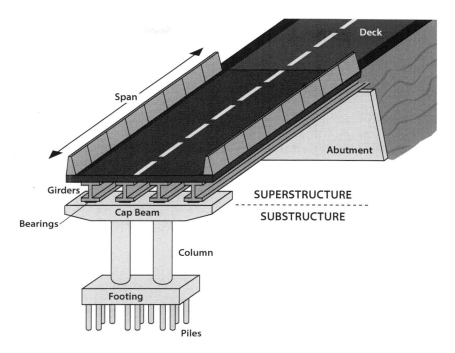

Figure 4.4. Girder bridge.

girder bridge, the deck-line coincides with its superstructure. The primary stresses that loads put upon it are bending and shear stresses.

Though the girder does withstand some compressive and twisting forces, its most distinctive structural role is to resist downward bending. Bending is greatest at the middle of the span, where *moment* (the leveraging effect of distance along the beam) is the greatest. Let us say we have in mind a steel I-girder. The longer it is, the more that a truck in the middle is going to make it bend down. For a longer girder span, the web has to be made deeper and the two flanges need a larger cross-sectional area. (Remember that the web is a metal plate that connects the upper and lower flanges and resists shear stresses.)

Made ever longer with the same cross-section size, the girder would eventually collapse under its own weight. If we proportionally increase the cross-section size, the bridge would cost too much. It is this fact that limits the lengths of girder bridges. A solution for reducing mass would be to put perforations in the web material, but that brings us to another type of bridge, the truss, to which we return later.

ARCH BRIDGES

Arch bridges may be classified into two subtypes, one being the traditional arch, which has its main structure below the deck line. Resting on the arch ribs called *spandrels*, the deck and its live loads exert downward compressive forces. Made of masonry or stone, the arched structure was the kind that the Romans were so adept at building. Some of them have withstood the ravages of traffic and weather for millennia. They have served so well because rock in itself is durable, large amounts may be obtained at low cost (no special metallurgy is required), and the material excels at withstanding the compression. Modern arches may, however, just as well be built of reinforced concrete or steel trusses.

The arch ribs transmit compressive forces downward and outward (figure 4.5) along the curvature of the arch itself. There should be very little tensile stress. If the arch is not to split apart at its center, the outward thrust must be met by a countervailing force. Rocky canyon walls fit the bill; they act as natural abutments. Where they are absent, enormous constructed buttresses will do the job, but at added cost.

Arch construction is an ancient art. In the traditional method, the bottom layers of blocks were piled on each other, building up the spandrels. As additional layers angled out over the span to be bridged, wooden scaffolding (called *centering* or *falsework*) held them up. The scaffolding continued to support the layers of blocks as they crept toward the center, until the

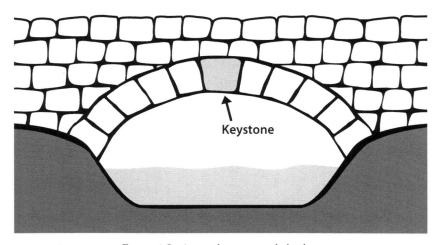

Figure 4.5. A simple stone arch bridge.

keystone was finally put in place. Since by now only compressive forces acted through the arch, the scaffold could be removed and the bridge would stay up. In modern construction, arch-shaped reinforced-concrete components are fabricated on land, lifted by cranes onto the scaffold, and then attached to each other. In some designs the actual arch is far lower than the deck. Vertical members called spandrel columns extend from the arch up to just-below deck level. At their tops, the columns are connected by horizontal beams, on which the deck rests (figure 4.6). In this design, the columns transmit loads downward to the arch, which then transmits them along its rib to the abutments.

The alternative subtype, known as the *through-arch*, has an arch-shaped superstructure, from which hangers extend down to the deck and hold it there by suspension. The deck exerts tensile forces on the hangers, which in turn apply downward forces on the arch-shaped superstructure, which then carries compressive forces outward and down to its abutment. When such bridges are meant to have a long span, old wooden scaffolding will no longer do and the arches become rather difficult to construct—over bodies of water, cranes may have to be placed precariously on barges.

TRUSS BRIDGES

A truss is often described as a lattice, lacework, or skeleton assembled from metal rods. It is an elegant and economical arrangement of materials applied to house roofs (these use wooden assemblies) and to the floors and roofs of high-rise buildings, as well as to bridges. Compared to the girder or arch, it uses less material for given span length.

Compared to the girder bridge, the truss bridge does require more labor and craftsmanship for assembling the structural members. As a span gets longer, a girder would have to get more massive and expensive, so the truss becomes increasingly attractive as the alternative, despite the higher

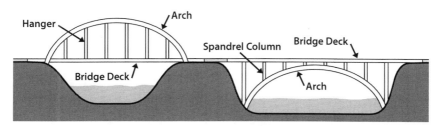

Figure 4.6. A through-arch and a deck arch.

cost of assembling it. The truss can even be thought of as a girder that has been enlarged and thoroughly perforated.

Viewed in cross section (you are standing on one of the approaches to the bridge, staring at the oncoming locomotive), the truss is a rectangle. Viewed from the side of the bridge, the members running along the top of the truss are known as the *upper chord,* and the ones along the bottom as the *lower chord.* The upper and lower chords are connected by a triangular pattern, since triangles are usually the most economical arrangement.

Several of these triangular patterns may be observed along a US road or railway, especially east of the Mississippi, where bridges tend to be older. The patterns on these older bridges have names, usually those of the engineers who invented them sometime during America's railway age, and there are people who pride themselves in remembering the names. The patterns are indeed interesting features of engineering history to look out for on a drive.

The *through-truss* is one on which the locomotive runs along the lower chord; a *deck-truss,* along the upper chord. Either way, the engineering principles are the same. As the locomotive is making its way along the truss, it curls the entire assembly slightly downward. The lower chord lengthens—it has come under tension. The upper chord shortens—it is under compression. The vertical and diagonal members at the sides of the truss manage shear stress.

A through-truss allows more clearance below. At some crossings, a combination works best: on approaches to the main span, the structure is a deck-truss; at the main span, it turns into a through-truss to provide more room underneath. As a passenger, you are first riding on top of the structure, then inside it. Most truss bridges have parallel upper and lower chords, but not all. Some are humped in the middle, where the structure's leveraging effect (the moment) is greatest. But this arrangement increases dead load in the middle of the span (figure 4.7).

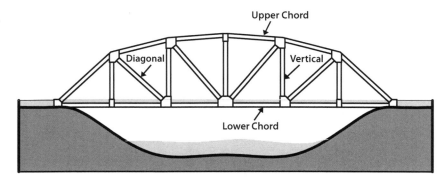

Figure 4.7. Truss bridge.

SUSPENSION BRIDGES

The immediately recognizable feature of the suspension bridge is the pair of cables extending from one side of a channel across two towers to the other side. Once a cable is suspended between the towers and anchored into the ground at both ends, it takes on its distinctive, graceful shape. Hangers extend downward from the cable to support the deck. Certainly the cable must be strong enough to support enormous loads. It consists of thousands of steel wires bound together, the whole cable then securely attached to an anchorage on each shore. The deck and its live loads exert tension through hangers to the main cables, which transfer forces to the abutments and the towers, which in turn transmit them to the ground (figure 4.8).

There is a debate, though not a particularly vociferous one, about the suspension cable's actual geometric shape. As long as the suspended cable is acted on only by its own weight, it has a *catenary* form, the form that any suspended rope or chain acquires. Once it has been loaded with hangers holding up the deck, the cable becomes distended, taking on a more or less parabolic shape, one group says. Then again, says another, it has loads suspended on hangers at regular intervals, creating slight changes in cable angle (or arc) between hangers, so the cable is more accurately described as having a *funicular* shape. The funicular shape will depend on how many hangers are used and how close the hangers are placed.

It is a geometric oddity that the shape of the main cable resembles an upside-down arch. Whereas the arch acts by compression thrusting outward along its ribs to the abutments, the cable acts by tension, pulling inward across the towers from the anchorages.

As its main span is supported only by cables, this arrangement is relatively light, but expensive to construct. The longer the span, the more financially viable it is. For spans of more than about 2500 feet, it is considered the only economical option. Its further advantage is that ground-

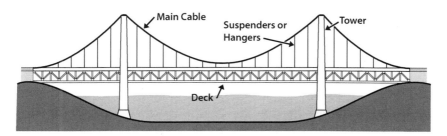

Figure 4.8. Suspension bridge.

level or marine construction has to occur mainly at the towers (and the anchorages). Once cables are draped, further construction can be done with workers and equipment riding on the cables. Construction at intermediate points on open waters can be avoided.

The longer the span, the taller the tower has to be to impart the appropriate shape to the suspended cable (to achieve economy). Towers often extend above deck line to heights three or four times that of the deck's distance to the land or water below. At the tower's crown the cables pass across a feature known as a saddle, where the cable can slide somewhat. Thereby the towers can be designed for the main job of resisting downward compressive force. Suspension bridges are especially susceptible to wind forces, so the deck is nowadays routinely built as a truss, to add stiffness.

CABLE-STAYED BRIDGES

Cable-stayed bridges are still relatively rare in the United States, but they are quickly increasing their share of medium-to-long-span bridges, so much so that the recent preference for this bridge type has been referred to as an infatuation. For many cities around the United States, the prospect of a new bridge raises hopes for a cable-stayed "signature bridge."

Like a suspension bridge, the cable-stayed bridge has its main structure above the deck line and supports loads by means of at least one tower. For a shorter cable-stayed bridge, one tower is sufficient. This is because structural members called stays extend diagonally from the tower to a girder, on which the deck rests. The stays transmit load forces by tension from the girder to the tower, which in turn transmits them to the ground (figure 4.9). As the stays are straight (whereas the suspension cable is curved), they provide a relatively stiff connection between tower and horizontal girder.

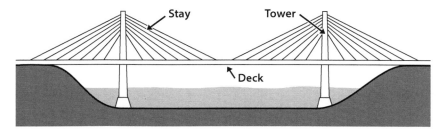

Figure 4.9. Cable-stayed bridge.

Towers differ in positioning and shape, as seen from the perspective of the motorist approaching the bridge. Some bridges have just one tower placed in the widthwise center of the deck (between opposite traffic lanes) or two towers each in the widthwise center but set apart longitudinally.

More commonly, two towers transversely straddle the deck. They may be connected by a horizontal beam on top, creating a rectangular form. Otherwise, they may angle inward, meeting at their tips, for a triangular shape; or angle inward, merge, and continue upward as a single pillar; or likewise angle inward but cross each other, making for a cruciform shape (figure 4.10). The choices are in significant part the architect's sculptural judgment.

For some spans for which a suspension bridge would need two towers, the cable-stayed bridge can perform well with one, making the latter more economical. The cable-stayed form is, therefore, used for intermediate-to-long spans, from 300 to about 1100 feet. Why aren't they as appropriate for the longest spans?

A basic reason is the angle by which the stay extends from the tower to the girder. To understand, let's note that there are two common arrangements of stays. In one, the stays all have the same angle and are parallel to each other. It is the other we now discuss. In this kind, the stay nearest the tower is angled very steeply downward (relative to the tower) while each additional stay is set at a progressively wider angle. The wider the angle, the less efficient the stay is at supporting the load of the deck and the traffic. Past 45 degrees, the stays pull more strongly on the deck, putting it under compression; as a countermeasure, engineers need to strengthen the deck, increasing cost.

Therefore, for very long spans, the tower has to be higher to permit the stays to be set at optimal angles. The stays must be correspondingly longer and must have more tensile strength. Alternatively, additional towers have to be built along the length of the bridge. Either way, the cost of the

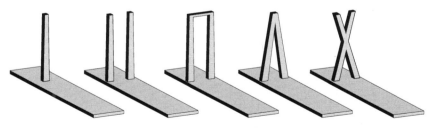

Figure 4.10. Tower shapes.

cable-stayed bridge escalates steeply as its span gets longer, to the point at which the suspension bridge becomes the more cost-effective choice.

As perceptive readers will already have noted, bridges do not have to fit neatly into the five types introduced here. An excellent example is Brooklyn Bridge in New York City. It is simultaneously a suspension bridge and a cable-stayed bridge, with some horizontal elements that are trusses, all held up by masonry towers. To have learned about girder bridges, arch bridges, truss bridges, etc., is just to have started entering the conversation. Bridge designers are welcome to mix and match, as long as they create a safe and cost-effective span, one that fits the requirements of the site and the community.

FITTING THE BRIDGE TO THE SITE

A basic criterion by which to select a bridge type for a site is the length to be spanned, but the matter is not as straightforward as it may first appear. For other than the longest spans, a number of bridge types can do the job. Any of four types of bridges might suit a 500-foot span. And even if we know the gap to be bridged, we may still have decisions to make about the length of the main span. A bridge could, after all, cross a 2000-foot-wide river with one span, two spans, or three or more all equally spaced, or with two multiple-span viaducts each extending 600 feet into the channel and then connected with a single 800-foot main span.

To be sure, community preference is always a prime consideration, but here we describe only technical factors, which citizens participating in bridge decisions should keep in mind. Span is one geometric criterion; others are clearances, angles of ascent and descent, and curves. Over water, the bridge may have to give clearance for tall ships; over a highway, for tall trucks. The need for curvature often restricts bridge type to girders, with box-girders preferred over I-girders, since the former are usually better at handling torsion forces imposed by traffic, wind, and earthquake.

For a long crossing that must be curved, one option is to have multiple girder spans near shore, each girder angled more than the previous so as to form a curve, after which the mainspan, which might be a through-arch, remains linear. If the deck must reach high above the surface, room is needed to bring ramps up to the proper elevation, so the geometries of road connections must be studied. Nearby roads are of special concern during construction, when bridge work can disrupt and endanger traffic, while the traffic itself can put construction workers at risk.

Geological conditions always matter. Rocky canyon walls and palisades might especially suit an arch bridge, providing it with natural abutments. As bridges are heavy, soils must be investigated for suitability as foundations. Depending on findings, the bridge piers may be made to rest on wide concrete platforms called *spread footings;* or on piles, which are posts driven into the ground; or in shafts drilled into the soil and filled with concrete (figure 4.11). If the soil is likely to settle, the bridge type selected must be one that can withstand some irregular settlement over time. The potential for earthquakes is always a concern.

Water is a concern and not just because of needed clearance. Rainfall and soil drainage affect the stability of embankments on which a highway bridge rests. Floods increase current pressure against piers or pose the danger the fast waters will scour out soil at the bottoms of piers and destabilize them. In designated flood plains, more piers may not be an option since they may obstruct drainage. And many environmental concerns come into play: preservation of natural shorelines, protection of endangered species, protection of wetlands, and others.

Let us say that planners call for a bridge made of a certain kind of concrete girder, but there are no nearby plants that make such girders. Yet there is a steel fabricating plant nearby. That might be a good reason to change the bridge design. Then again, let us say that the construction is to occur in a narrow gully to which large prefabricated parts are difficult to deliver. Then it might be better to assemble a truss bridge from steel

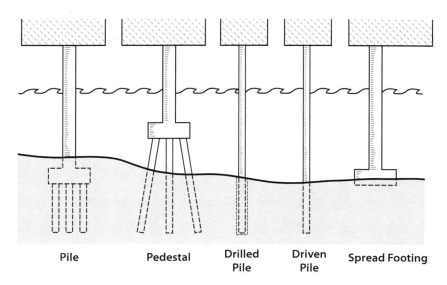

| Pile | Pedestal | Drilled Pile | Driven Pile | Spread Footing |

Figure 4.11. Piers—footings and foundations.

elements, which are smaller and easier to transport. In short, nearness of industry, nearness of labor skills, and construction-site features can all affect bridge selection.

Traffic projections, maintenance methods over time, hazard prevention (as from coastal surge due to hurricane), and esthetic judgment all count. Most of all, there is always the master consideration, cost. To give just one rule of thumb: adding one more span to a multispan bridge adds one more pier plus the work of attaching the added girder. It is usually more economical to build with one fewer span and instead to lengthen each girder along the bridge. But the longer girders then have to be made with deeper webs and heavier flanges to resist bending forces.

Asked whether to rehabilitate a bridge, replace it, or build a new one, decision makers must contend with just these kinds of complex choices and tradeoffs. The two major considerations are cost and safety. Before we get to analyzing costs, we should recognize one more matter, one that nonengineers may have difficulty appreciating. It is that engineers do not have perfect knowledge of how a bridge will perform. In deciding the kind of bridge that will fulfill travel needs at least cost, the engineers face uncertainties about future live loads and the proposed bridge's capacity to resist them.

Sources and Further Reading

In the advanced engineering text *Design of Highway Bridges: An LRFD Approach*, 2nd ed. (Hoboken, NJ: John Wiley, 2007), authors Richard M. Barker and Jay A. Puckett also classify bridge types and discuss the selection of bridges to suit site conditions. Brian Hayes provides an interesting guide to bridge types and other man-made scenery in *Infrastructure: The Book of Everything in the Industrial Landscape* (New York: W.W. Norton, 2005). Another useful, though dated, source is M. S. Troitsky, *Planning and Design of Bridges* (New York: John Wiley, 1994).

FIVE

MAKING STRONG BRIDGES

‗‗‗‗‗‗‗‗‗‗‗‗‗‗‗‗‗‗‗‗‗‗‗‗‗‗‗‗‗‗‗‗‗‗

Dealing with Uncertainty

When a city or state sets out to build a bridge, several motives may be at work. It may want to relieve congestion, complete a highway that has a missing link across a river, provide traffic access to a new office complex, or replace a faulty old bridge. And it may in addition want to get construction jobs and create a civic object that serves as a lasting monument. Especially for unique or spectacular ("signature") bridges, the path from idea to steel-and-concrete edifice brings members of numerous professions into play. They include everyone from architects and financial analysts, geologists and project managers, and drilling technologists and lawyers, to transportation planners to project car use and geotechnical engineers to make sure that soil conditions can support the structure.

In all bridges long or short, magnificent or ordinary, engineers retain fundamental responsibility for strength and safety. We have already met (in chapter 3) some of the concepts the bridge engineer works with, including stress and strain due to compression, tension, bending, shear, and torsion. This engineer must make sure that the bridge structure can resist the stresses that loads impose on it. It is on accurate calculation of loads and resistance that the structure's safety depends. So it may come as a surprise that engineers must depend for these calculations on information that contains plenty of uncertainties.

Yes, they can turn to the grand analytical tradition of applied Newtonian physics, the results of generations of professional experience, and the outcomes of engineering research and testing. Yet the uncertainties remain, both about future loads that a planned bridge will have to carry and about the future performance of structural members (and the entire bridge) under those loads. Since about the year 2000, most engineers in the United States have adopted a new method of bridge design. It is the method professional engineering societies and many states now require—a method that recognizes that, inevitably, uncertainties must be taken into consideration in design.

The method is called *Load and Resistance Factor Design,* or LRFD for short. Preceding a fuller explanation, it is worth saying that the method (1) adjusts load by uncertainty factors and (2) adjusts resistance by such factors; hence the F in LRFD is the abbreviation for *factor.* (Engineers in Europe use a similar method under a different name.) Participants in bridge decisions should understand why this is the guiding method for bridge engineering: why it serves to increase bridge safety.

To architects, *design* refers to the imaginative creation of the overall structure, with a view to safety and cost for sure, but also to its fitness to place and esthetic qualities as experienced by users. To bridge engineers, design (sometimes known as *engineering design*) has a more specific meaning. It is the selection, arrangement, placement, and assembly of material and components into a system that provides sufficient safety and serviceability at acceptable cost. The challenge to the engineer is to know which system provides just that.

Our purpose in the chapter is to explain the reasoning through which the engineer accomplishes this design task. For those who will follow this explanation, there is an important proviso. The vast majority of bridges are of short-to-medium spans, so the engineer does not need to work through the entire design logic explained here. Rather, he relies on professional codes and standards, which specify the components, sizes, and materials appropriate for normal circumstances. Even for these routine bridges, the professional committees that drafted the codes had to base them on the design principles (as well as research results and professional experience) to be described here.

SETTING LIMIT STATES

The perceptive reader will say that this is now abundantly clear and hardly worth saying: London Bridge may be falling down, but the bridge we design certainly shouldn't. Our engineer should make sure that loads do not exceed the structural limits at which damage occurs.

But our problem is not that simple. Everyone agrees that the bridge must not collapse, but it may well become damaged enough to be unusable or rickety long before collapsing. Or it may behave well enough for a while, but undergo deformations and metal fatigue after two decades of stresses. So the engineer has to decide the kind of damage he is trying to avert. The upper threshold of resistance against which he sets projected loads depends on the kind of damage being averted.

Engineers refer to these damage thresholds as *limit states*. Through guidelines developed by professional standards committees, bridge engineers must (with some exceptions) make bridges conform to these defined limit states. The American Association of State Highway and Transportation Officials (AASHTO), the primary standards developer for bridge engineers, identifies the limit states as *strength, service, fatigue,* and *extreme event.* As the strength threshold is the one we will consider in this chapter, we summarize the others just briefly.

The *service* limit refers to the quality of the bridge for its users. If in the course of use the bridge is excessively bouncy, vibrates too much, causes too much jarring, otherwise provides an unpleasant or frightening ride, reduces traffic speed, and damage vehicles, then it is not properly providing its intended service. The bridge may also crack at multiple points, tilt or dip in a way that makes the driving surface uneven or causes traffic accidents, or vary from rough to smooth to slanted to bouncy along different segments. The bridge may exhibit these inferior forms of service quality while still posing no danger of collapse. When designing a bridge, the engineer must as one of his duties ensure that bridge performance does not fall below standard limits of acceptable service quality.

The *fatigue and fracture* limit state refers to potential damage from persistent cycles of traffic loads. During years of traffic, structural components are subjected to repeated stresses, which may initiate and then over time exacerbate cracks, deformations, fractures, and damage to connectors. The cumulative damage reduces service quality and can even result in catastrophic fractures, which undermine bridge strength, leading to collapse. As fatigue occurs from stresses applied repeatedly over long durations, it requires special methods of analysis.

The *extreme event* limit state seeks the structure's survival during earthquake, severe collisions, ice flows, and scour at the bridge foundations. We return to this subject in the next chapter.

Now we get to the threshold of acceptability on which we concentrate for the rest of the chapter. It is the one most people would expect to be the determining factor in bridge engineering: the *strength* limit state. For a structural member, say a girder, strength is the stress it can undergo without fracture or other serious damage threatening collapse: the greater the strength, the higher the stress the structural member can withstand. The strength limit of a cable in tension is determined in part by the elastic limit, beyond which the cable stretches precipitously for added units of load.

For a bridge assembly as a whole, the strength limit state is the standard for the highest combination of loads the structure can withstand while retaining its structural integrity. The strength limit state is set at the threshold beyond which the structure undergoes distress, structural integrity is impaired, and repair is required (but collapse is not imminent).

CHECKING FOR STRESS MAXIMA

To find out whether a structure does or does not exceed its limit state, engineers must perform a remarkably large number of calculations. For a girder bridge, they must determine the future strengths of girders, piers, abutments, and connections (such as bolts) under the loads that will be placed on them.

To illustrate what the engineer must do to compute the various stresses as against the strain limits, let us consider a 400-foot three-span steel girder bridge supported on abutments at each end and two sets of two-column concrete piers, as shown in figures 5.1A and B. Note that each pier consists

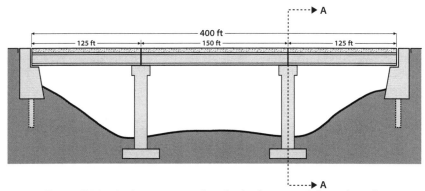

Figure 5.1A. A three-span steel girder bridge viewed from the side.

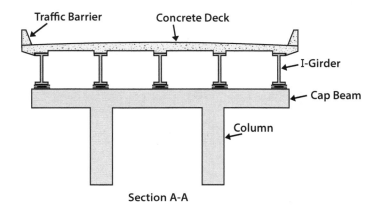

Figure 5.1B. The same bridge viewed in cross section, revealing a two-column pier, the columns connected with a cap beam, on which five girders rest, supporting the deck.

of two columns connected to each other on top by a *cap beam*. Composed of five parallel girders, each taking the "I" shape, the superstructure supports a deck four lanes wide.

Let us confine ourselves just to the measurement of load and resistance on one girder—the middle girder over the middle span. Since stresses will vary along the length of the girder, we will seek the stress maxima—the points along the girder at which stress is greatest under given loads.

We begin by measuring the expected dead load, for which we shall consider just the weight of the girder, and that of the concrete deck, parapet (railings or wall along the sides), and light fixtures and any other utilities on the structure. Given the length of the girder and the dimension of its cross section, we can readily check dead weight in a professional manual, such as that published by AASHTO, from which we may project a dead load per linear foot of traffic lane.

Recall that the girder must first of all resist its own dead load. Even at this bridge's modest main span, the girders will sag an inch or so in the middle.

Now we go on to vehicular live load. The standard practice is to estimate the effect of routine car traffic plus one heavy truck over a particular linear foot of deck. From studies that have been conducted, routine traffic is estimated to exert 0.64 kips per linear foot per lane. Professional standards also provide a model truck (figure 5.2), which consists of a cab weighing 8 kips over the axle, plus a truck bed carrying a container.

So far, we have assumed that the vehicular live load is static—as during a traffic jam. Now we include the fact that it is likely to be moving, and thereby bumping up and down, exerting additional downward loads. To account for this dynamic load, AASHTO requires an additional allowance of 33 percent above the normal vehicular load and the truck axle loads.

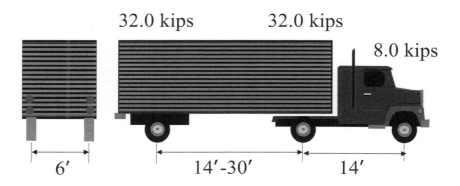

Figure 5.2. A model truck for estimating bridge loads.

There is an additional allowance for braking forces, but since these are longitudinal, they do not affect the vertical resistance of the bridge. Here we are looking only at vertical effects. We have left much else out as well. Pedestrian load is excluded because our bridge has no sidewalks. Waterway effects are excluded. Special effects of any bridge curvature are no complicating factor, since our bridge is straight. Thankfully, we do not get into these complications here.

Given these simplified assumptions, the engineer has to estimate points of maximal stress. Typically, shear stresses are highest where the girder intersects the abutment and the pier—and these shear stresses will be larger or smaller depending on the position of the model truck on the span. At any point along the girder, the truck will exert a downward force, pushing the girder down. The bending exerts maximum tension effects on the bottom flange of the I-beam, maximum compression on the top flange, and shear effects in the web. Bending moment is typically greater the farther we are from the supports, so it is typically highest in the middle of the girder.

Upon such calculations, the engineer can discover the points along the span of maximal shear stress and maximal bending stress. Then, she can determine whether the I-girder can at these points resist these stresses. The resistance depends on the construction material, the dimensions of components, and how they are connected. Steel has fairly consistent qualities, so engineers can consult tables of materials and cross-section properties to help them estimate whether the expected loads would damage a girder of given dimensions.

Let's say our study shows that the girder we have chosen as an example can indeed resist the projected loads at maximum points of stress. The engineer's job is far from done. There are many more contingencies to be considered. What if the two piers, which have been driven into the riverbed, eventually undergo settling, burrowing down by two inches? Now the girders they support will sag two inches more than originally calculated. How will this affect the girder's ability to resist the given truck loads at the points of highest stress? More calculation is needed.

To be sure, the engineer could have recommended a girder so thick that it could have held up against almost any conceivable load, even several tanks rolling on it at once. But that may not to be a good solution. For one thing, the girder's dead weight would grow, requiring piers with greater carrying capacity, and increasing downward pressure, forcing greater downward settlement on the pier. What is more, the cost would escalate. As much as it always seem to be an unmitigated good thing to increase safety, we must remember that ever-increased safety for one bridge deducts from the amount to be invested in other bridges, or in other public concerns, whether highway safety or public health.

So the engineer's job is to keep the bridge cost to the amount needed to make it strong enough to meet anticipated loads, plus some safety factors to account for unanticipated stresses. That brings up the next problem: just how is the proper safety factor determined?

THE OLD METHOD: SAFETY FACTORS

The loads are often designated by the letter Q, for no apparent reason other than that this letter may be underused elsewhere in engineering. To remember that Q means "load," think of "quartz," which indeed makes for a rather heavy load, close to 20 times heavier than an equal volume of cork. Resistance is simply designated with the letter R. For a safe bridge, the effects of load must be less than resistance, or in short: $Q<R$.

The engineer would be pretty foolish to make the girders just barely strong enough to support typical truck and car traffic. For one thing, the load (Q) that the bridge will carry cannot be foreseen with certainty. It is possible that on a winter day, during rush hour, a convoy of trucks carrying nothing but quartz tries to cross the bridge, while the bridge is weighed down with ice.

What is more, even the bridge's resistive capacity (R) is uncertain. The steel girders can vary in their quality because of variations originating at the foundry, or may vary in the effectiveness of connections at the joints.

So the prudent engineer designs the bridge to account for possible excess in loads or shortfalls in resistance. In short, he incorporates a *safety factor*, F, in his design, altering the basic formula to $FQ<R$. For example, he may want that even twice the expected peak load should be less than expected resistance, hence $2Q<R$. The safety factor here is two.

In simplified form, we have introduced the venerable design doctrine called *Allowable Stress Design*. The doctrine can be traced back to the early part of the nineteenth century. For each component being evaluated, the engineer calculates applied stress against the allowable stress. A respectable engineering method, it served well for over a hundred years, and with progressive improvements, was proper procedure for bridge design until the late twentieth century.

But it did leave open an important problem: that of knowing how to set "F." Set too low, it poses dangers; set too high, it raises costs beyond what is necessary. The level at which F is set should also depend on structural redundancy in the bridge—the capacity in the bridge to have some components take up the load when other components have failed.

Through their professional associations, bridge designers have developed the practice of setting up codes committees, which undertake (or commission) the studies through which to set required safety factors. These studies involve careful examination of historical records on bridge loads, such as vehicular load, wind pressure, snow and ice, and earthquake. Over the years it has become clear that the magnitudes, frequencies, and distributions of these loads vary greatly, and that mathematical approaches based on probability would be the best ways of assessing risks and setting safety factors. It has been with the intent of putting the calculation of risks on a more rational and consistent basis that new design doctrines have evolved.

THE CURRENT METHOD: LOAD AND RESISTANCE FACTOR DESIGN

We skip a step in the evolution of bridge design doctrines and move right into the current structural design philosophy, the LRFD, which uses probability curves to depict both load *and* resistance. This may seem odd. It is obvious enough that traffic, weather, and other loads vary, and so probabilistic thinking helps depict the frequencies of rare but dangerous combinations of loads. But the bridge structure has a durable existence with supposedly preset resistive qualities, so why treat resistance probabilistically?

As it turns out, the properties of materials also vary. Looking at a girder of given material and size, we are not perfectly sure how well it resists loads. Concrete of given dimensions may vary in its properties because of quality control at the fabrication site, differences in temperature and humidity during drying, and quality of workmanship during on-site assembly. Though modern manufacturers go to great effort to achieve consistency in the structural materials they produce, microscopic examination of even steel components reveals tiny fractures and imperfections, which cannot be fully eliminated.

In a large sample of a specific type of structural units (whether girders, cables, slabs, struts, columns or joints), the units vary from being below standard, to standard, to above standard. Because of the efforts to maintain uniformity in fabrication, the vast majority of units will have standard strength, and a few will have more than the required strength, but a few others will be below strength. This result can be depicted graphically as a normal curve.

Now observe figure 5.3, which shows both the load curve (Q) and resistance curve (R) on the same graph. In this graph, the vertical axis refers to probability per lifetime of the span. The horizontal axis refers to load intensity, measured in kips. As the structure's resistance should be greater than the loads it is expected to support, the resistance curve is to the right on the horizontal axis. The next observation is very important: the two curves overlap, if only slightly. Under the overlapping area, from just under the intersection point to the left, load exceeds resistance. This part of the overlap is the danger zone: it depicts the rare but possible situ-

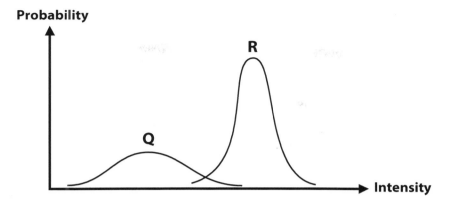

Figure 5.3. Load (Q) and resistance (R) curves, shown on the same plot.

ation in which loads exceed the strength limit state and threaten severe damage. Note that this comparison of the curves on the same plot can be made only when they fulfill specific mathematical conditions (namely that they are "normal curves").

Under the same mathematical conditions, we can create a combined curve that represents the remainder when load curve is subtracted from resistance curve. The resulting R-Q curve is shown in figure 5.4. Now the shaded area to the left of the Y-axis shows where load exceeds resistance—it

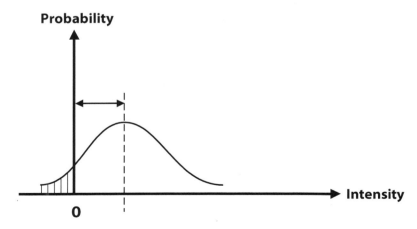

Figure 5.4. Reliability curve, also known as the *failure probability curve*. Failure probability is shown in the striped area to the left of the Y-axis.

is the danger zone. The design engineer must strive to minimize this danger zone while keeping in mind the cost of increasing resistance capacity. Earlier we asked how we set the safety factor that separates load from resistance. The LRFD method provides a standard, measurable way to make a choice.

Let us say that committees of engineers and experts determine that a bridge is acceptable if it has no more than a .135 percent (.00135) chance of failure (from dead load plus live load) in its lifetime. This means that no more than .135 percent of the area under the resistance curve should be in the danger zone. If figure 5.4 were correctly illustrating this example, the striped area to the left of the Y-axis would take up .00135 of the area under the whole curve. This is known as the bridge's *failure probability*. Put positively, there should be a .99865 chance that the bridge will not exceed its strength limit state over its planned lifetime—this is known as its *reliability*.

This reliability (or failure probability) standard is applied not just to the bridge as a whole but to each load-bearing bridge component. Relying on this standard, the practicing engineer can, in principle, figure out whether a proposed bridge is meeting safety standards or not.

A WORD ON REDUNDANCY

We have given only the briefest sketch of the LRFD method of bridge design. Among the many topics we have left out is the effect of structural redundancy on the reliability of the bridge.

Let us consider a suspension bridge holding up the deck on a series of hangers (cables extending vertically from the suspension cable to the deck). Since many hangers hold up the deck, an accident in which a truck severs a hanger does not pose a disastrous risk. After all, the other hangers offer redundancy: loss of one or even two hangers would not compromise the bridge. Therefore, the reliability of the hanger could be set lower than for components that are nonredundant. For example, the suspension cable itself is nonredundant. Severe damage to even one of the two suspension cables could be catastrophic. Therefore the reliability requirements for the suspension cable must be especially stringent.

BRIDGES: SAFER BUT NOT PERFECTLY SAFE

Now let us recap the advantages of the LRFD approach. First, it sets a reliability factor that each component as well as the entire assembled bridge system must meet. Second, it accounts for variability in both load and resistance. Third, it uses scientific testing and professional standards committees to assess the reliability of most of the kinds of structural members that the engineer is likely to use in a bridge. (For bridges having exotic designs with unusual structural members, the engineers will have to commission original testing.) Thereby, the engineers can achieve a standard level of safety among bridges.

In the historically used Allowable Stress Design method, in which safety factors were largely based on rules of thumb derived from experience, engineers were under great pressure to make safety factors high, which of course increased cost. The modern engineer following the LRFD method can apply more logical and general principles for assessing reliability, and so may have the confidence to make the safety factor smaller, especially for loads and resistances on which there is good scientific knowledge. Present-day bridge engineers have more confidence in the safety of their designs.

The new method forces all of us to realize that even bridge engineering, a well developed discipline based on scientific research, is subject to uncertainties, and that a small percentage of bridge components in the United States will fail each year, and very rarely, an entire bridge. Deep within the engineering discipline, design depends on a value judgment on just what level of failure is acceptable. The LRFD method faces a particular challenge with respect to uncertainties that are hard to quantify and express, as because of lack of data or lack of sufficient scientific understanding. This problem comes up most of all with respect to the dangers posed by extreme events. We discuss the problem in the next chapter.

Further Reading

Richard M. Barker and Jay A. Puckett, *Design of Highway Bridges: An LRFD Approach*, 2nd ed. (Hoboken, NJ: John Wiley, 2007). American Association of State Highway and Transportation Officials, *LRFD Bridge Design Specifications*, 6th Edition (2012). M. Myint Lwin, "Why AASHTO LRFD?" *Transportation Research Record* 1688, Paper No. 99-0935 (1999).

SIX

RESISTING EXTREME EVENTS

Anyone who crosses a modern American bridge should be assured that, even if it is long and slender, and spans choppy waters in a deep chasm, it is likely to be among the safest of technological creations. Of 600,000 or so bridges in the United States, only about 30 to 40 (less than .01 percent) a year "fail."

To structurally "fail" doesn't necessarily mean to collapse. It more likely means that the structure's use has to be suddenly reduced or terminated—not just briefly interrupted—because it poses immediate structural danger. In keeping with strict inspection standards established by the US government, highway agencies are well equipped to identify such perilous bridges. Sometimes for years in a row there are no casualties in America from bridge collapses. Among the safest things one can do in this country is cross a bridge.

That our bridges so rarely fail reflects well on the state of the engineer's art and on the organizations, whether construction companies or governments agencies, that build and maintain them. But when catastrophic failures do occur, they do cause personal tragedy and, for the areas they serve, great economic loss. Bridges are often bottlenecks for traffic, so when a bridge becomes inoperable, the regional economy suffers the loss of supplies, employees, and the ability to send out its goods—much more than for the loss of a length of road.

In decision making about a future bridge, the structure's reliability and its ability to sustain life safely must have a central place. The awareness of safety must include the construction phase, when accidents are most likely because of human error and the dangers that accompany the use of complex construction equipment on narrow ledges, exposed to the wind. This chapter will, however, disregard the construction risks. We look instead at the hazards that threaten a bridge in the course of its operating life.

In a typical sequence, engineers first balance potential loads against resistance to ensure that the bridge meets requirements for the *strength limit state*: that it will perform well under its dead load and range of normal live loads expected during its lifetime. Once they ascertain that it will, they can get on to the next big question, the *extreme event limit state*. They ask: over the same lifetime, will this bridge also perform well in the most intense earthquake shaking (or other dangerous force) likely to impact it? But before we say more, we should demonstrate that extreme events are indeed the most important dangers to a well constructed bridge.

BRIDGE DISASTERS

As of this writing, the Minneapolis bridge disaster of 2007 is still on many peoples' minds. On August 1 of that year, during rush hour, the eight-lane Interstate 35W bridge suddenly collapsed, killing 13, injuring over 100, and removing an economically crucial Mississippi River crossing. In April of the same year, in Oakland, California, on an elevated interchange approaching the Bay Bridge, a gasoline tanker crashed and overturned, spilling gasoline that erupted into flames, causing heat so intense that the elevated structure melted and collapsed. Fortunately, it was the middle of the night and the only person injured was the driver. More recently, on May 23, 2013, an oversized truck struck one of the upper truss members of a through-truss bridge over the Skagit River in Washington State, leading to one span's collapse. The collapse fortunately caused no fatalities, only injuries, but it did sever a critical highway link between Seattle and Vancouver, disrupting traffic on a wide scale.

A bridge disaster during the early morning of September 15, 2001, was largely ignored by the national media because it came just days after the 9/11 attacks. It occurred in southernmost Texas, near the Mexican border, at the Queen Isabella Causeway, which joins the mainland to South Padre Island. Four loaded barges crashed into a support column, causing two spans, and later a third, to plunge into the channel. The collapse occurred near the highest point of the bridge; approaching drivers could not see the gap, and several cars fell into the water, killing eight.

Over a decade earlier, when the Loma Prieta earthquake struck California on November 17, 1989, Oakland was served on its waterfront by a two-level highway, each having four lanes. During intense shaking, a mile-long stretch of the upper level collapsed, falling onto the lower deck and crushing or trapping cars below, killing 35.

One more sad event will suffice for now. In April 1987, heavy rain and floodwater, including spring snowmelt, washed away (the precise term is *scoured)* the soils under the Schoharie River Bridge under the New York State Thruway, upstate New York's most heavily travelled highway. After enough soil had washed away, the piers progressively collapsed, causing the entire bridge to fall. Unsuspecting motorists drove into the gap; ten bodies were recovered.

These are very different kinds of bridge failures. The Minneapolis collapse differs from the others, because it was attributable—investigators found—to a design failure. The alleged flaw was in the gusset plates, which are metal objects that connect truss members to each other. With bolts, rivets, or welding, the metal plates are meant to firmly secure the joints between structural members. With added stress from construction equipment parked on the bridge, the plates reportedly failed, dooming the bridge. The primary problem was internal to the bridge design itself.

In the other five disasters, however, the collapse is primarily attributed to an event external to the bridge structure: a tanker crash and resulting fire, truck impact, barge impact, earthquake, and scour. These are what engineers refer to as *extreme events*.

THE CAUSES OF BRIDGE FAILURE

We have to be careful about bridge "failure" statistics, in part because data on the subject are not uniformly or comprehensively collected. According to the National Bridge Inventory data provided in chapter 2, as many as 100,000 bridges are structurally deficient or obsolete. But their inadequacies are well understood and do not pose immediate danger. Some undergo planned closing after inspectors determine them to be unsafe, but even they are not immediately dangerous. They have, rather, simply served out their expected service lives or are being met with far heavier loads than they were designed for; they have not "failed" in the ordinary sense of the word.

Where actual failure does occur, the causes are often difficult to diagnose. Failure occurs under complex interaction between internal factors (inadequate structural resistance) and external ones (loads exceeding the bridge's design load).

Sources of internal failure may include faulty engineering design, flawed detailing documents (submitted by contractors with engineers' approval), error or malfeasance during construction, or materials deficiencies. The internal failure may also be traced to deterioration due to age and to inadequate inspection and maintenance. Otherwise, failures are externally caused: by loads greater than ones that the bridge was expected to resist.

When a bridge fails, the judgment about what caused it to do so should be made through investigation by qualified forensic engineers. They examine original design and detailing documents, review inspection reports, and build computer models that replicate the process of collapse. It is the forensic study that should say whether the failed bridge did or did not fulfill building codes and expectations of the profession; whether it had been maintained well enough; and whether bridge problems should have been recognized and the bridge closed. The study should judge whether unexpected or extremely rare external events disastrously stressed a bridge that was, by accepted criteria, well built.

Without consistent forensic studies, we usually cannot know with confidence what makes American bridges fail. Researchers have nonetheless been able to review the evidence and come to plausible conclusions about the most common causes of failure.

The study we have in mind (by one of us, George Lee, along with colleagues Satish B. Mohan, Chao Huang, and Bastam N. Fard) found evidence on 1062 bridge failures over 33 years, for an average of 32 per year (table 6.1). "Failure" was here defined as collapse, partial collapse, or distress sufficient to force reduction in traffic, including closure.

The authors attributed some failures to internal problems, including deterioration, corrosion, construction problems, and faulty design. But they found that the large majority of failures, altogether 88 percent, had external causes (table 6.1). Most of these were extreme events, the subject to which we now turn.

Table 6.1. Causes of Bridge Failure, United States, 1980–2012

Cause			Percentage				
			2000–2012	1990–2000	1980–1990		
External causes	Hydraulic	Flood	4%	15%	10%	28%	47%
		Scour	5%	9%	5%	19%	
	Collision		5%	5%	5%	15%	88%
	Overload		3%	3%	6%	13%	
	Fire		1%	1%	1%	3%	
	Earthquake		0%	1%	1%	2%	
	Wind		1%	1%	0%	2%	
	Environmental Degradation		2%	2%	2%	7%	
Internal causes	Design, Construction, Material, etc.		3%	3%	5%	11%	11%
Other & Misc.			—	—	—	—	1%
Total			24%	39%	36%	—	100%

Note: Total bridges judged to have "failed" over 33 years: 1062
Source: Lee, Mohan, Huang and Fard, 2013.

PLANNING TO AVERT DISASTERS

As the table shows, flooding is the most common cause of failure and is most numerous in years in which there are large regional floods, impacting numerous bridges. That was certainly true of 1993, when Midwestern floods ruined many.

It was true again in 2005, during the Katrina disaster and subsequent hurricanes. In hurricanes, it is generally not the winds that damage the bridge but the surge from the ocean, which in turn pushes waters upstream. At the same time, hurricane rains drench entire watersheds, from which floodwaters pour into rivers. In 2005, several bridges from New Orleans to areas north of Lake Pontchartrain underwent catastrophic failure, as did bridges in Mississippi and Alabama.

Floods are indeed a large danger, but they may still not be uppermost in bridge planners' minds. River floods (but not flash floods on steep streams!) occur with enough early warning that bridges can be closed before there is collapse and loss of life, though with the accompanying harm that evacuation from nearby affected communities may be hampered. Viewed over a larger period of time than that examined by the researchers, earthquakes also cause much bridge damage, in the rare years in which large-magnitude quakes occur. Earthquakes (and collisions) cause damage essentially without warning, and therefore pose greater danger to life.

For planners and engineers (and for the professional standards committees that guide engineers), the tough problem, therefore, is to decide which events are likely enough that it is worth building the structure to resist them. To decide, we should forecast the *probability* of the extreme event and its *intensity*, which can be a measure of wave force, or ground shaking, or collision force. Put differently, we have to forecast the probability that the hazard intensity will exceed some limit during a time period, such as the bridge's expected life span.

What is more, we have to weigh the cost of mitigating the hazard against the benefits in safety to be gained. If a very intense vessel collision or earthquake has only a one-in-ten-thousand chance of occurring per year, then perhaps it is not worth investing in countermeasures. This is not because we are heartless, but because it may be much more cost-effective to seek safety elsewhere, as by reducing traffic accidents, which kill far more people than structural collapses do. As we shall see, these are tough decisions. They pose engineering and planning dilemmas to this day.

EARTHQUAKE

The United States undergoes a large regional earthquake only once every decade or two, but when it does occur, it can damage multiple bridges. As compared to floods, earthquakes threaten even bridges, such as elevated highways, over dry land. And an earthquake occurs at most with a few seconds' warning, which is not enough to vacate a bridge.

Earthquakes pose great challenges to the specialized professionals, known as earthquake engineers, who design structures to lessen their effects. For one thing, an earthquake can shake a bridge up and down, or transversely (at a horizontal angle to the direction of traffic), or longitudinally (in line with traffic). The shaking can also soften the soils on which the abutments and piers rest or in which the foundations are buried, destabilizing the structure's connection to the ground.

What is more, any single type of shaking has differing effects on various bridge components. A pier that withstands up-and-down shaking may do less well when pushed in a forward-backward direction in line with bridge alignment. The problem becomes all the more difficult if the bridge is curved. Whereas much of traditional engineering concentrates on static structural members and the static loads they must support, the earthquake engineer must consider a complex dynamic structure stressed by dynamic loads.

Most of us are used to hearing earthquakes measured in terms of the Richter scale. It measures the energy released at the location underground where the quake is centered. If we draw concentric circles over the land surface radiating from the epicenter and measure ground shaking, we see that the effect tapers off with distance, but not uniformly. Water bodies and rock strata vary the intensity of shaking at points around each of the circles. For the engineer, the pertinent question is not Richter scale magnitude at the earthquake origin, but the direction (up and down, transverse, longitudinal) and intensity of shaking at the bridge itself.

The local intensity of the quake is best measured in terms of ground acceleration. Why we say ground *acceleration* instead of ground movement is readily understood. If you place a toy structure on your palm and carry it at a constant velocity, it will stay in place. But if you suddenly speed up the movement of your hand (or suddenly slow it down), the toy will topple over. Similarly, what affects the bridge is not the mere movement of the earth, but acceleration—a change in velocity.

For the engineer, usually the most pertinent measure of earthquake intensity is *peak ground acceleration*, meaning the highest acceleration the ground undergoes during the quake, as measured in *g*, the metric also used for airplane acceleration. Bridges in major regions of the United States are expected to resist earthquake shaking of .4g—though there is much variability in this requirement. Other important measures besides ground acceleration include movement direction, frequency and duration of ground shaking, and any ground displacement.

Of directions of shaking, the longitudinal direction tends to be the most dangerous. This stands to reason. Under static conditions, the structure is primarily meant to resist downward forces of the superstructure, and since the bridge is longer than wide, to resist tipping toward one side or another. But longitudinal shaking imposes forces that the bridge may be least prepared to resist, as seen in figure 6.1.

What can the earthquake engineer do? One thing she cannot do is reduce the intensity of the quake force. No one can stop these enormous geological forces. One option is to strengthen resistance by making the structure stronger so it can resist earthquake force; another is to make the foundation and footings stronger. These are expensive options that may nonetheless prove inadequate in the face of the tectonic energy nature has in store.

Earthquake engineers have developed additional countermeasures, among which *energy dissipation* devices are especially sophisticated. To see how they work, let us imagine a north-south girder bridge subjected to longitudinal shaking, in which one phase of the shaking moves the girders in a northward direction (the opposite phase moves it back southward).

Anticipating such an event, engineers attach a *damper* connecting the girder to the bridge's north abutment. The damper is a large cylinder containing a piston filled with viscous fluid. (A viscous fluid, like honey as compared to water, is thick and sticky and more resistant to stresses that would make it flow.) The piston rod is attached to the abutment and the cylinder to the underside of the girder (figure 6.2). When the northward movement occurs, the girder suddenly shifts toward the abutment, squeezing the piston into the cylinder. The viscous, incompressible substance in the cylinder cushions the movement of the piston. As a result, the girder accelerates less than the ground does—the damper *dissipates* the earthquake energy, mitigating harm to the structure.

Alternatively, engineers can install a *bearing* between the girder and the column or abutment on which it rests. Before we describe bearings,

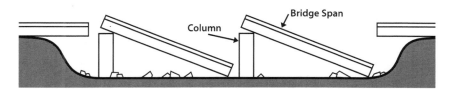

Figure 6.1. Bridge span unseated by longitudinal shaking.

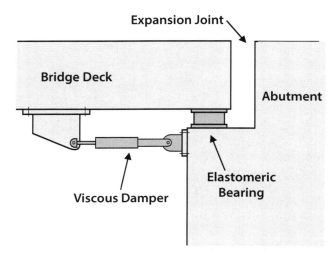

Figure 6.2. A viscous damper and elastomeric bearing connecting a girder bridge to its abutment.

let us pause for a minute to explain that some bearings have ordinary uses unconnected to extreme events. Even without regard to earthquake threat, engineers must install bearings between the superstructure and the substructure. In a girder bridge, the bearings transmit the weight of the bridge span and vehicle loads to the column, which in turn transmits it to the ground. Some bearings help reduce traffic vibration and allow expansion or contraction from heat variations.

Now let us return to earthquake regions: here the bearing has to take on additional duty. During an earthquake, it must structurally isolate the girder from the abutment on which it rests. This is called a *seismic isolation bearing*.

Imagine the bearing as a box connecting the underside of the girder to the abutment. In one example, called an elastomeric bearing, the box is flexible or rubbery. As the ground moves northward, it deforms the bearing, stretching it in the northward direction. The acceleration of the girder is reduced accordingly. The bridge is then partly isolated from the impact.

Alternatively, the bearing can consist of two plates, one atop the other, separated by steel rollers that allow sliding. The upper plate is connected to the girder and the lower one to the abutment. When the ground moves northward, the upper plate will not move as much as the lower plate. The rollers reduce the bridge's impact against the abutment.

Whatever the countermeasure, whether increased structural strength or energy dissipation with a damper, it adds to the cost of the bridge. The more intense the anticipated ground shaking, the more to be paid for countermeasures. But when earthquakes are studied around the world, it turns out, as one would expect, that the most intense ones are also the rarest. Against the largest but least likely, it may simply not be worth investing in countermeasures. The decision on how much to invest must be made in part according to the importance of the bridge.

A structure may carry so much traffic and be so essential to the economies it connects (especially in areas where the highway system has little or no redundancy—no other nearby highway bridge can take up the traffic pressure) that planners deem it to be a *critical bridge*. By professional standards, critical bridges must be considerably more hazard-resistant than ordinary bridges.

To know just how earthquake resistant to make the bridge, engineers must be given the peak acceleration, such as .4g. One way to set this ground acceleration value would be to estimate the greatest ground acceleration the bridge may expect in its lifetime—say it is just 75 years. According to AASHTO specifications for regular bridges, that number should reflect the most intense earthquake that could likely recur every 375 years—five times the actual 75-year expected life of the bridge. For important bridges, the specifications call for a larger peak acceleration, one that would only occur every 1000 years. For even more critical bridges, the bridge may be designed to resist the most intense acceleration likely to occur in 2500 years.

HYDRAULIC FORCES FROM FLOOD

A large proportion of US bridges (listed in the National Bridge Inventory) cross bodies of water, so water hazards must indeed occupy the bridge planner's attention. On steep and narrow rivers, a particular danger is flash flood, in which upstream rains and snowmelts rapidly drench narrow and steep channels, causing sudden rises in water level. These are the floods that can damage bridges with little warning.

Under widespread heavy rainfall covering many Midwestern states, the region's large rivers swell more slowly, giving time for warnings and closures of bridges. In recent decades, the rivers appear to have become more susceptible to floods because of changes in land use. As land is covered with impervious surfaces (parking lots, roofs, roads), water no longer seeps through the soil but rather flows rapidly across the land surface. Rainwater also runs rapidly through cities because it is channeled through storm sewers. On rivers, embankments and levies meant to protect upstream towns increase water velocity toward downstream settlements. Taken together, these may be some of the reasons for the increased severity of Midwestern floods in recent decades.

As we saw after the Katrina disaster, coastal floods from hurricanes can cause huge destruction. On water bodies near New Orleans, floodwaters reached all the way up to the bridge decks. Bridges are designed to resist water levels reaching part way up the piers. When the water extends up the full height of the bridge, the structure has to resist not just the full brunt of the moving water, but also the water's upward buoyant forces, tending to force the span upward. Then floating logs, boats, shipping containers, vehicles, and large appliances smash against the bridge. (Not least, the moving water may scour away the soils at footings and foundations, but we leave that topic for later.)

As with earthquakes, no single metric sufficiently measures flood intensity. A fairly good single measure would be wave force and velocity. AASHTO has only in recent years begun to establish threshold wave force and velocity criteria for bridges.

An option unavailable for earthquakes may now be feasible for floods: to lessen the likelihood of a flood occurring in the first place. Through proper regional planning that manages the placing of impervious surfaces on land, authorities can regulate water flows over land so as to reduce flood risk.

Where floods cannot be stopped, engineers can opt to strengthen the bridge: design it with deeper and thicker foundations, strengthen the piers and pier-to-superstructure connections, build the superstructure at higher elevation, or shelter bridge piers with rock piles and artificial islands.

Once again, the operative question is how much to invest in counter-measures. To answer, engineers have to agree on (their professional associations must set standards for) the flood threshold that the bridge must be designed to resist. It may be, as earthquakes are, expressed as the largest event likely to occur in 100 or 500 years.

HYDRAULIC FORCES FROM SCOUR

Scour refers to the process that washes away sand and rock from the stream-bed, leaving spaces or gaps that can destabilize the supports on which a bridge rests. Scour can be dangerous even if it happens gradually over the years, as normal river flow stirs up and suspends sediment and transports it downstream, while the process is hidden from view. It is more dangerous in raging streams, because swift water has more energy by which to lift and carry the sediment. Scour can also be uneven. Rivers that meander may erode soils at higher rates on one side of the river than the other, exposing some piers to more scouring than others. The very existence of a bridge can add to the problem, because the piers and abutments reduce the area through which the water flows, increasing water velocity (figure 6.3).

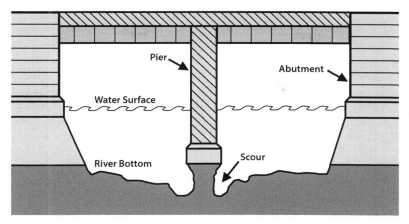

Figure 6.3. Scouring at a bridge. Image adapted from US Geological Service publication.

This process affects the bridge rather differently than other hazardous forces we discuss in this chapter. Earthquake and wave forces endanger a bridge by increasing the loads impinging on it. But scour causes harm by reducing the structure's capacity to resist forces. This happens as the flowing water wears away the soils at the bridge footing and the underlying foundation (figure 6.4).

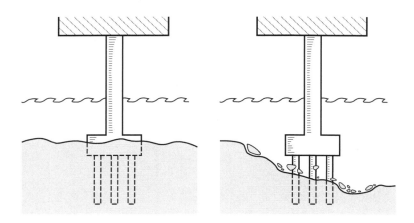

Figure 6.4. Scour wears away soils at the bridge foundation.

A footing once buried in soil is now exposed to moving water, in effect making the exposed pier taller than it once was. The lengthened pier must resist wave forces greater than it was designed for. It is then that collapse can ensue.

To judge how much to spend on bridges to avert scour, we need a measure of scour intensities, and no single measure of scour intensity will do. Among the contenders are measures of the lowering of the streambed, contraction of the stream cross section, and changes in flow, such as flow acceleration or appearance of flow vortexes at piers.

The chance of scour can be reduced. Regional planners can help reduce it through upstream land-use changes that reduce the rate of inflow from snowmelt and rainfall. Environmental agencies can conduct periodic stream inspections and clear away ice dams and debris. Where scouring remains a risk, engineers can opt to strengthen bridge footings and foundations. Or they may surround the foundation with *caissons*. Sometimes made of concrete, these are cylindrical shapes embedded into the stream bottom to dissipate the stream's power of washing away soil near the piers. Or again, engineers may reinforce the streambed, shoreline, foundation, and abutments with piles of rock called *riprap* (figure 6.5).

A state transportation department may build caissons or riprap at a bridge after the adjacent soil reveals the occurrence of scour. But it is risky to wait until the scour actually becomes visible. Ideally, the agency

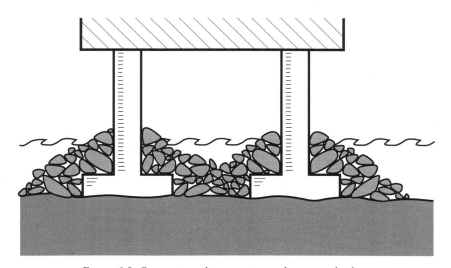

Figure 6.5. Riprap to reduce erosion and scour at bridge.

should reinforce all susceptible bridges. But the agency may own thousands of bridges, subjected to varying scour risks, and it is just too expensive to reinforce all of them.

One recommendation is that the threshold scour should be set at the highest intensity likely to be encountered over 75 years, the typical design life span of a modern bridge. Like earthquake and flood hazard criteria, this one too depends on accuracy of available data, this time data on stream behavior. But stream scour is a slow and complex phenomenon, for which scientific forecasts may be very uncertain. With new technologies, it is increasingly affordable to install monitoring systems at bridge piers. These measure scour depth and give warning when action must be taken against scour.

VESSEL COLLISION

Ocean-going vessels have become so large that, even at slow speeds, their momentum makes it impossible for them to suddenly turn or stop. In America and abroad, collision with bridge piers has indeed caused catastrophic collapse with loss of life.

The destructiveness of the impact depends on the ship's mass, speed, and kind of impact, whether direct frontal hit by the bow, sideswipe by the hull, or impact by the deckhouse against the span. It depends also on the ship's structural deformation (absorption of energy) upon impact as well as on the bridge components' capacity to deform or become displaced without collapsing (figure 6.6).

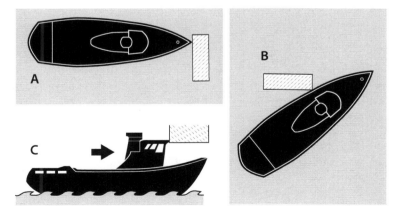

Figure 6.6. Ways a vessel can collide with a bridge. Adapted from L. Gucma, "Methods of Ship-Bridge Collision Safety Evaluation," 2009.

Among countermeasures, it is best to altogether prevent the collision by keeping the ship accurately in navigable lane (the *fairway*), as with electronic guidance systems and training for tugboat operators. But weather alters currents, sandbanks shift, wayward boats force other vessels to take evasive moves, barges are sometimes improperly loaded, and every once in a long while a pilot is less than sober. So, we could reinforce the bridge with stronger piers and pier foundations, though that's an expensive solution and one by no means foolproof against gargantuan ships.

It may be cheaper to add a fender to a pier. If we are speaking of a small bridge threatened by small vessels, the fender could be nothing more than a rubber skirt draped around the column. Larger vessels must be met with massive fenders built of logs or concrete. Or again, ships could be deflected with clusters of logs fixed into the riverbed, or even with artificial islands.

As always, the question is just how much it is worth investing in such measures. The decision should depend on frequency of ship traffic, ship sizes and draught, water depth, width of the fairway, and navigational complexity (currents, winds, turns, underwater obstacles). The more critical the bridge—traffic, economic impact, availability or lack of alternative crossings—the more to be spent.

How intense a blow should the bridge (and its protective fenders) be designed to avert? How improbable should that event be before planners decide that it is just not worth trying to avert it?

The 2009 AASHTO specifications provide a standard. For "critical bridges," the expected annual frequency of a collision severe enough to cause collapse should be no greater than .0001 (once in 10,000 years). For regular bridges, the expected annual frequency should be no greater than .001 (once in 1000 years).

TERRORISM

It has been speculated since 9/11 that terrorists might target bridges. By comparison to natural and accidental hazards, which hold no preference for famous bridges, terrorists would likely focus on long-span bridges that are national symbols and have great economic importance.

In one respect, an attack against a bridge can be less threatening to life and limb than attack against a building: users of the bridge are traversing the structure and not (as in a building) occupying it. If the bridge and its approaches are not congested with traffic, then the mere suspension of vehicle entry will quickly clear the bridge. Bridge landings provide easier and faster egress than do the doors to a high-rise building. With proper surveillance, bridges can be closed early in response to suspicious activity.

In other respects, bridges are more vulnerable than buildings. Bridges are subject to the constant flow of vehicles, which as a practical matter cannot be inspected for explosives. Bridges expose their structural members for all to see, so a malicious but trained observer could perhaps estimate the vulnerable points at which explosives could catastrophically harm the structure. And long-span bridges generally have less redundancy than buildings do. Some cable-supported bridges may be especially vulnerable, though cables are generally thick enough and well-enough anchored that it would take much determination to damage them. Cable-supported bridges are also some of the most architecturally dramatic and so could attract attackers searching for high-visibility targets.

Several news reports have suggested that bridges have been targeted by terrorists, but such allegations range from the well documented to mere rumor and unfounded suspicion. Potential threats have caused controversy for the Brooklyn Bridge, the Chesapeake Bay Bridge, Golden Gate Bridge, San Francisco-Oakland Bay Bridge, and the five-mile long Mackinac Bridge in Michigan.

Let us assume that terrorists will attack only with explosives placed on the deck, cable, or pier—we disregard missile attack or purposeful ramming by a ship loaded with explosives. But how powerful a bomb could an attacker assemble? Might it be an explosively formed penetrating device that does more damage (as compared to a conventional device of the same charge) in the direction in which it penetrates? What about the intensity of fire that follows? Looking for an answer, an engineer could propose a scenario for a bomb that an ordinary van could carry, and then design standoff distances to piers to be protected from the bomb. Recurrence intervals (annual probabilities) are anyone's guess and are probably unknowable.

THE STATE OF THE ART

In a typical bridge design procedure, engineers carry out calculations on a proposed bridge type to ascertain that it is strong enough to resist its own dead weight and the live loads to be expected in its service life. Having checked against strength limit states, they must also do so for extreme-event limit states. They seek guidance from professional standards committees, which periodically issue updated requirements. These requirements are enforced by state transportation agencies. But even if enforcement is weak, and if morals of ensuring life safety do not stir them enough, engineers are well advised to pay attention to standards. If for no other reason, they would want to preserve their licenses and maintain some protection against lawsuits.

Most of all because they are educated to be good professionals, engineers as a natural and proper part of their routine do extreme-event checks. Their challenge is that the standards are themselves in question.

Against earthquake, though specialists readily agree that further research and refinement are needed, the shaking-limit states are relatively well established, with adjustment by US region, according to seismicity. For hydraulic hazards, mainly flood or scour, each river carries a different risk. Standards may be seen as less urgent, since river flood generally occurs slowly enough that bridges can be evacuated so that, even if the bridge is lost, human life isn't.

Against vessel collision, limit states must be set based on local estimates of collision intensity and probability. Perhaps because the professional committees setting collision standards have sat separately from those setting earthquake and scour standards, the collision recurrence probability for ordinary bridges is, as of this writing, more stringent. Then again, there is good reason to be extra cautious, because collision can cause immediate collapse.

Against terrorist bombs on bridges, there is no standard, for several reasons: no US bridges have been bombed, we cannot estimate the probability of such an event occurring, and we would have to guess at the blast intensity. Research is being conducted, but as of yet there are no formal guidelines.

Overall, though American bridges are indeed safe as compared to almost all other technologies, the bridge-engineering community still has considerable room to improve its safety record, while keeping the price affordable.

Answers are hard to come by in part because it is not at all easy to estimate how an extreme event will affect a future bridge. Even when we assume a certain collision intensity or blast intensity, engineers may still be unsure how a particular bridge will react—whether it will successfully resist the extreme force or will not. Under normal daily stresses, the interactions among bridge components are already a complex matter; they are all the more so under extreme events. Engineers might simply be unsure how the bridge system as a whole will behave at such times.

Even as we write, there is progress on several fronts. There is growing professional awareness of the need for consistent treatment of risks across hazard types. Data on hazard propensities is accumulating. Statistically inclined engineers conduct research on better ways of modeling risks. They face a particular hurdle: current practice has different approaches to the four limit states: strength, service, fatigue, and extreme event. Ideally, the engineer should be able to combine at least strength limit calculations and extreme-event calculations within the same probabilistic framework. For advancing the state of the art in bridge engineering, such conceptual progress will be a key issue.

For studies of the effect of earthquake, engineering research teams place a bridge prototype on a *shaking table*, a large platform that subjects physical model structures to shaking (up and down, sideways) under progressively greater accelerations. Engineers working on flood and scour conduct similar reduced-scale tests in hydrological test beds. After actual bridge failures, forensic teams investigate the process by which failure occurred and report back to the profession.

Most expensively, in just a handful of field laboratories in the world, full-sized structures are put under destructive testing—actual bridges are destroyed—to examine the precise dynamics by which they fail. Finally, a foretaste of the future: new structural health monitoring systems embed monitors and sensors into the structure itself. These emit flows of data that describe how the bridge behaves under the various stresses to which it is exposed, giving engineers new tools by which to assess how hazards impact bridges. To better adapt bridge costs to safety needs, a long-run ideal for engineering research is to equalize bridge-failure probabilities (reliability standards) across hazard types.

Of the great Roman arch bridges built over two thousand years ago, only a few have survived. Many of the rest were destroyed by a type of extreme event, namely earthquake. If we want to build bridges that will last long, our biggest challenge is to make them survive extreme events. As US infrastructure ages and bridges have to be replaced, the opportunity arises to apply new technologies and new insights, to build new generations of bridges that are safer and more durable than ever before.

Further reading

"Bridge Failures—A Summary and Evaluation" by Lubin Gao, Ph.D., P.E., and Richard A. Lawrie, FACI, P.E., presented at New York City Bridge Conference, 2007. Michel Ghosn, Fred Moses, and Jian Wang, "Highway Bridge Design for Extreme Events," National Cooperative Highway Research Program Report 489, Transportation Research Board, National Academy Press, Washington DC, 2003. George C. Lee, Satish B. Mohan, Chao Huang, and Bastam N. Fard, "A Study of U.S. Bridge Failures (1980–2012)," MCEER Technical Report No. 13-0008, University at Buffalo, June 2013. For detailed technical information, see *AASHTO Guide Specification for LRFD Seismic Bridge Design*, 1st Edition, 2009; *AASHTO Guide Specification and Commentary for Vessel Collision Design of Highway Bridges*, 2nd Edition, 2009. *AASHTO LRFD Bridge Design Specification*, 6th Edition, 2012; and *AASHTO Guide Specifications for Bridges Vulnerable to Coastal Storms*, 2008.

PART III

BRIDGE PLANNING

SEVEN

IS IT WORTH IT?

Costs, Benefits, and Tough Decisions

Whether to build an item of infrastructure, or reconstruct it, or find ways to avoid the need for it—these are tough decisions in part because they are expensive. While humanity does have many vices, the lack of imagination about how to spend money is not one of them. Should the money go for a four-lane downtown truss bridge with room for light-rail cars on the assumption that public transit will increase, or a six-lane suburban cable-stayed bridge where traffic has been growing and land is cheaper? Then again, shouldn't it rather go for a new airport, waterfront park, or hospital emergency rooms?

The bridge project will require years of engineering design, environmental studies, and construction, and will have to be paid for before its benefits are felt, so there are likely to be shorter-term needs on peoples' minds. The money could, after all, be used to fix potholes, which may be doing more to slow traffic than the lack of a bridge does, or could be just the ticket for bridging not the local river but the year's budget gap. So, when it comes to, say, a few tens of millions set aside for a new bridge, it's safe to say that there will be contending ideas about what to do with the money. Of several ways of trying to resolve the issue, the most widely respected is cost-benefit analysis, which turns out to be a fairly intricate procedure.

It is not its logic that is difficult; the fundamental ideas are straightforward. Rather, the problem is that, to do the analysis, we have to make lots of assumptions, sometimes guesses, and the data we gather may be wrong. What's more, we may have to decide whether the bridge seems worthwhile before, paradoxically, we have done all the necessary studies. It would, after all, be easier to make the decision once full engineering design has been completed, along with geological studies so we can assess the site's geological hazards, and traffic studies to see how the bridge fits into the regional road network. But these studies are in themselves expensive—often 10 percent or more of construction costs—so we would like a first estimate of the bridge's worth. It is that sort of first estimate, a preliminary cost-benefit analysis, that this chapter will illustrate, along with the many conceptual traps into which the analysis can fall.

Let's have in mind a city bisected at its downtown by a river 900 feet wide. The two bridges that presently exist are congested for hours around morning and evening rush hours, backing up traffic into downtown, and worsening the traffic jams downtown. So there is a proposal on the table: build a new four-lane bridge. New bridges at places that didn't previously have one are becoming rarer in the United States, as the construction of the nation's road systems has leveled off. Most new bridges are replacements for old ones, which have to be demolished, often with the enormous complication that traffic has to be served even while the new bridge is being built. For the sake of having a clear example, so our case is simpler: a brand new bridge where none existed before.

For preliminary analysis (because full engineering has not yet been done), it is foreseen as a deck-arch bridge with footings resting on rock near the river edges. The arch must be high enough to allow tall-mast boats to pass underneath, so the arch supports a deck that is higher than the roads at shore level. Total deck length will be 1200 feet, longer than the arch span, to give room at the approaches and ramps to reach the requisite deck height. The deck width is 50 feet, giving room for four traffic lanes and a pedestrian lane. The bridge is to be built entirely on public property, so no land acquisition will be required—a very important cost savings. However, extensive ramp work is needed to connect the bridge to the city's roads and highways. Should the bridge be built?

MEASURING INTERNAL COSTS AND BENEFITS

An essential though incomplete way to answer the question is by comparing the proposed bridge's costs and benefits. For now, we mean *internal* costs and *internal* benefits. These accrue only to the bridge project itself and leave out, say, the noise and exhaust the bridge creates or the possibility that quicker mobility will spur investment in new office buildings downtown. In studying just the internal effects, we confine ourselves to the narrow world in which all we care about is cost to the bridge owner-operator (that's the city government) and benefit to the bridge users. We get only later in this chapter to the problem of *external* costs and benefits, which affect the city, the region, or society as a whole.

Even for the seemingly straightforward question "How much will it cost to build?" the answer is, during early estimation, far from certain. Construction costing is difficult. There is site work with excavation, backfill, and the driving of piles; tons of concrete for the deck and arches plus steel for reinforcement; and various kinds of labor needs, not to mention the construction equipment, from concrete mixers to truck-trailers, cranes, and crawler tractors. And there is the problem of labor and equipment productivity: just how much can they accomplish per hour in the river mud and in the fog that often envelops the downtown? The answers are, furthermore, subject to price changes after the estimate is made, and not just because of overall inflation, but also because costs will fluctuate item by item, depending on supply and demand.

For a shortcut, we can just estimate costs of a bridge by looking at the state's average construction costs per square foot of deck area. These are quite reliable for short highway overpasses and can give a reasonable range with which to start, but are less reliable for predicting the unique specifics at a longer river crossing. Having to make an early case on whether the bridge makes sense or not, the initial estimator has the further problem that the engineering and environmental studies, which are still ongoing, may conclude that the structure has to be different than first expected. They may show that rocks meant to support the feet of the arches are less reliable than expected, requiring deeper drilling, hence greater cost. Or seismic study can reveal a greater-than-expected chance of earthquakes, forcing a redesign that makes the bridge more resilient during shaking but also makes it more expensive.

Maintenance cost estimates offer a particular opportunity to make the bridge seem cheaper than it is. The authorities making the decision can easily be lulled into letting that happen. They are under pressure to solve downtown traffic problems at the river's edge, but also under pressure to reduce city and state borrowing. They want to borrow just enough to solve the problem. When built, the bridge will be new; deferred maintenance won't be noticed for years. Under pressure for results now, the authorities may not want to be bothered with the price a decade from now of repaving the deck that weathered too quickly or fixing a pier that was not built strongly enough to withstand minor boat collisions. By constructing the bridge with less durable materials now, it is possible to shift costs to those who will later be responsible for maintenance. To prevent such tricks, infrastructure decision makers must remember that the correct desideratum is the *life-cycle cost*, which includes the original project cost plus all costs to accrue over the years of service.

Measuring the costs is difficult enough. The measurement of benefits is even more so, because the methods are more susceptible to error and there is much legitimate disagreement over what a correct measure would be. Somehow the correct method would capture the benefit from a new convenient route across the river: the benefit of avoided travel time. As we shall see in the next chapter, it is possible to project how travel patterns would change with the new bridge, perhaps reducing travel delays. Among the many difficulties of such models is that a new, less congested road in itself gives an incentive to more people to drive—so the very existence of the bridge (not counting any increase in business activity, population, and employment) might put more vehicles on the city's roads than were there before. However difficult the effort may be, the model should be able to measure the avoided travel time as the bridge's primary benefit.

What adds greatly to the travel modeler's burden is the second step in which she wants to figure out how much this avoided travel time is worth. The answer includes the operating cost of the vehicle: gasoline, insurance, and depreciation. It also includes the cost of the travelers' time, perhaps at their average salary rate, though of course such rates vary greatly. To answer, the analyst may well rely on standardized data, like that for Minnesota in table 7.1.

Table 7.1. Recommended Standard Values for Vehicle Operation, State of Minnesota

Operating Costs	$/mile
Auto	$0.23
Truck	$0.62

Value of Time	$/person hour
Auto	$10.46
Truck	$19.39

Crash Values	$/crash
Fatal	$3,600,000
Injury Type A only	$280,000
Injury Type B only	$61,000
Injury Type C only	$30,000
Property damage only	$4,400

Source: Minnesota Department of Transportation, Benefit Cost Analysis Guidance, June 2005

Since shorter travel times reduce accidents, the conscientious mod-eler will also take a third step—that of estimating the value of avoided accidents. In view of current state accident rates, the analyst can estimate the reductions the new bridge will bring, and estimate averages for avoided vehicle damage (repair or replacement cost) and injuries (health care and lost income costs). Minnesota's transportation department recommends that accidents be broken up into three classes of severity, each of which has a cost attributed to it (table 7.1). Accidents also cause deaths, to which it is con-ventional to attribute a life cost, in Minnesota's case $3.6 million in 2005, though perhaps inflation has since increased Minnesotans' nominal value.

The idea of attributing dollar value is likely to alarm those who first encounter it. Should we let the Minnesotan die if the cost of saving him is $4 million? Should children be valued less than adults, who are income earners, or more than old people, who have less left to live? Despite first impressions, however, modelers we have known have not been ghoulish. They're concerned, after all, about an abstract calculating standard called a statistical life. And they are, let credit be given, confronting a difficult problem in transportation planning: roads *do* generate accidents and deaths, while they also provide access and mobility, which are essential to life. By building a stretch of roads to a certain construction standard, which will inevitably pose a given statistical risk of future fatalities, and not spending the additional million for the finest paving and roadside barriers, the planner does implicitly give a monetary valuation to lives of future users, whether he admits it or not.

Now let us recap on the measurement of bridge benefit. We obtain, first, the number of travel hours that the new bridge allows vehicles to avoid; second, the average vehicle operating costs and traveler time value per avoided hour; and third, the average dollar value of avoided accidents per hour. Their product is a measure of bridge benefit.

The first step is clearly essential, but the second and third steps are worth questioning. While they get closer to the full benefit against which to weigh cost of the bridge, they pile ever more measurement assumptions on top of data that is already shaky. New inaccuracies compound old inaccura-cies, possibly making the complex answer less reliable than the basic answer.

It is even questionable whether accident data should be included at all. Adult travelers implicitly accept risk when they travel; that risk is incorporated into their valuation of whether they should travel or not. In America, highways and streets are subject to laws, liability requirements, and professional standards. If a road exposes travelers to a significantly greater risk than is standard, then the authorities have the obligation to fix the problem, by repairing it, slowing traffic, or closing it. Unless a road is being planned specifically to replace an unsafe road (greater road safety is a specific objective), it makes sense to assume, in accounting for benefits, that the new road's safety benefits are simply proportional to travel time. So for a preliminary decision, it can be adequate just to complete the first of the three steps—to find only the cost per avoided travel hour.

For the sake of illustration, here are two simplified decision rules. First option: divide total bridge cost by the number of avoided travel hours. If cost per avoided travel hour is low compared to other, similar bridge projects, then the new bridge appears worthwhile. Second simple option: compare total bridge cost with total bridge revenues that travelers would reasonably be expected to pay through tolls. If tolls will pay for bridge engineering, construction, maintenance, and upkeep over the bridge lifetime, we have good grounds for saying it is worthwhile. But remember that these must be preliminary results, because we are so far considering only *internal* costs and benefits.

SETTING UP THE ANALYSIS

In our example, we take the second simple option, the comparison between bridge cost and toll revenue. But even this method will turn out to have its challenges.

To follow our analysis, look at the cost column in table 7.2, where we have removed three zeroes from every column—you have to multiply the figures by a thousand in your mind. The costs of our hypothetical bridge are shown over three years of construction and a 30-year service life. Costs over that time are given in constant dollars—we assume zero inflation. Many would rightly disagree with us about the 30-year assumption (why shouldn't it be built to last 50 years?), but please give us a pass on that for now, if only to keep us from squeezing a longer table onto the page.

As the table shows, the cost of building the bridge will be $33 million, one third to be expended during each year of construction. But of that, only $18 million goes for actual construction. Another $9 million is spent on engineering and architectural design, traffic studies, soil studies, environmental studies, legal fees, and inspections. And $6 million goes for on-shore road and ramp reconstruction. For those who need to make the public decision, it is of course total project cost that matters, not just the construction cost.

Once built, the bridge needs to be operated and maintained for its 30 years. Keeping to our assumption that there is no inflation, we estimate annual preventive maintenance, administration, safety patrols, and toll operations at $400,000, an amount considered respectable for keeping the bridge in good shape in the long run. An additional $3 million is needed every twelve years for repairs and repainting.

Table 7.2. Costs and Benefits of a New Bridge in Constant $1000

Year	Cost	Benefit	Net Cash Flow	PV @ 3.5%	PV @ 5.0%
1	−11000		−11000	−10628	−10476
2	−11000		−11000	−10269	−9977
3	−11000		−11000	−9921	−9502
4	−400	2500	2100	1830	1728
5	−400	2500	2100	1768	1645
6	−400	2500	2100	1708	1567
7	−400	2500	2100	1651	1492
8	−400	2500	2100	1595	1421
9	−400	2500	2100	1541	1354
10	−400	2500	2100	1489	1289
11	−400	2500	2100	1438	1228
12	−400	2500	2100	1390	1169
13	−400	2500	2100	1343	1114
14	−400	2500	2100	1297	1061
15	−3400	2500	−900	−537	−433
16	−400	2500	2100	1211	962
17	−400	2500	2100	1170	916
18	−400	2500	2100	1131	873
19	−400	2500	2100	1092	831
20	−400	2500	2100	1055	791
21	−400	2500	2100	1020	754
22	−400	2500	2100	985	718
23	−400	2500	2100	952	684
24	−400	2500	2100	920	651
25	−400	2500	2100	889	620
26	−400	2500	2100	859	591
27	−3400	2500	−900	−356	−241
28	−400	2500	2100	801	536
29	−400	2500	2100	774	510
30	−400	2500	2100	748	486
31	−400	2500	2100	723	463
32	−400	2500	2100	698	441
33	−400	2500	2100	675	420
Sum	−51000	75000	24000	**1042**	**−4316**

30-Year Service Life   Discount Rate at 3.5% or 5%   PV=Present Value

Note 1: This is an arch bridge of 1200 ft deck length, 50 ft width. Deck area is 60,000 sq ft. Construction cost estimated at $300 per square foot. Project cost (construction plus all studies, permits, and inspections) is 50% more: $450 per square foot. Construction takes place over first three years. Additional $6 million over the same three years for on-shore ramp and road work.

Note 2: The present during which the decision is made to build or not build would be Year 0. Costs and benefits are accrued at the end of each year. For example, the entire construction expenditure in Year 1 is assumed to have occurred on Dec. 31.

Note 3: Annual administration, preventive maintenance, operations, and toll collection costs $400,000.

Note 4: Costs after construction are for operations and maintenance. Rehabilitation occurs every 12 years, costing $3 million each time.

Note 5: Benefits consist of toll revenue of $1 per crossing, average of 6850 crossings per day, 365 days a year, equaling $2.5 million per year.

Note 6: The two bottom-right cells, shown in bold, are **Net Present Value**.

The toll receipts (see the benefits column in table 7.2) start arriving in the fourth year, the year in which the bridge opens. Since the bridge is urban and operating at moderate speeds, we're projecting that, during a peak traffic hour, about 1500 cars cross. The rest of the day has fewer cars, adding up to an estimated daily average of 6850 (the average includes the lower traffic on weekends and holidays). At a one-dollar toll for this many vehicles (with no special estimate made for trucks and buses) over 365 days, the annual revenue is $2.5 million, an amount assumed to stay identical for 30 years.

The net annual cash flow, shown in the third column, is then the sum of costs and benefits, which turns out to be negative during construction years and during the 15th and 25th years (repair and repainting), but positive, reflecting a healthy toll income, for the other 28 years. Prospects for our bridge look pretty good, as long as we remember that estimates are just that—they're estimates meant for discussion, questioning, and revision.

THE DISCOUNT RATE AND PRESENT VALUE

Questions of whether to build or not to build public works are capital investment issues. Private firms face similar investment problems when they have to decide whether to build a building, buy a machine, or start a new product line. A distinguishing feature that makes them "capital investment" is that decisions must be made about the asset now, though costs, benefits, and risks from that asset flow back over years. For our arch bridge investment, they flow back over three years of construction and thirty years of service.

In the United States, expensive transportation infrastructure projects are usually paid for, in large part, with state and federal funds. But, to keep things simple, we will assume that our hypothetical city itself pays for the bridge. So our mayor has in front of him a bridge-financing problem. The year's tax receipts have to pay for regular expenditures, like police protection and street cleaning, not to mention his salary. It is not fair to ask residents to try to pay for the bridge from current tax revenues, since people including those yet unborn will be using the bridge for decades. It makes sense to borrow the money and pay for it over the structure's life span. Fairness aside, the year's municipal budget has nowhere near enough money for a bridge anyway.

The mayor now has to decide (that's another simplification, as if city councilors and various interest groups were pawns in his hands) whether to go ahead and borrow money for the prompt start of construction next January 1. He inspects table 7.2 but only takes the time to look at the columns labeled "costs," "benefits," and "net cash flow." "This is a great project!" he says. Total project costs plus operations and maintenance over 30 years add up to $51 million, but toll revenues to $75 million, with net revenue of $24 million. Is he mistaken in thinking that this is such a fantastic project? Mistaken indeed, because he has forgotten the time value of money.

After all, the city has to borrow money and will owe interest on it. (Or even if it had the money in cash to start with, it could have put it in a safe interest-bearing bond. By investing the cash in the bridge, it is losing a stream of bond interest.) Let's just consider Year 4, when the bridge will have net cash flow of $2.1 million, which we can call FV for future value. This future amount is worth less to the city now in Year 0, the decision time. This is true even if the $2.1 million receipt four years from now is guaranteed—if there is no risk about the city's receiving the funds. Say the city now has a spare $1,830,029 in the coffers. The city could invest that exact amount now in a bond at say 3.5 percent interest, yielding $2.1 million by the end of the fourth year. It follows (if the toll revenue is just as secure as the bond—an important proviso, but one on which we don't elaborate here) that the fourth-year toll revenue has *present value* equal to that of the bond just mentioned. There is then a way of figuring out the equivalence between present value and future value, given a *discount rate*. The discount rate can often be interpreted as the "interest rate," but not always, for reasons to which we return below.

Let us refer to the fourth-year revenue of $2.1 million as the future value (FV). We can get its present value (PV) with the following formula: $PV=FV/(1+r)^t$. In the formula, r refers to the discount rate. Let's say it's 3.5 percent (that is to say r=.035). The t refers to the number of years (in this case four). Applied to our problem, the present value of the fourth year's revenue is 2,100,000/(1.035 X 1.035 X 1.035 X 1.035), yielding $1,830,029, as shown in the table. That of course equals the present cost of a bond that will pay an annual 3.5 percent over four years.

In our table, if you follow down the column for the 3.5 percent discount rate, you will observe that for each year in which there is net cash flow of $2.1 million, the present value progressively diminishes, until in the 33rd year it's worth only $674,820. With discounting done for each year at the 3.5 percent rate, the net present value (commonly known as the NPV) of all net cash flows is just over $1 million.

So here is the lesson for the mayor: if he takes into account the time value of money, the bridge is still a good investment—it still generates more benefit than cost—but is not as great as he first thought. When economic conditions are riskier, lenders may charge higher interest, even if inflation expectations are still held at zero. To take account of that possibility, we also consider the building of our bridge under the discount rate of 5 percent (in the final column in table 7.2). Now, cash revenues of $2.1 million in the project's fourth year have even less present (Year 0) value than before, and progressively less in each future year. The project's NPV now falls into the negative: the present value of the cost now exceeds the present value of the benefit by over $4 million. (See the bottom of the rightmost data column in table 7.2.) The mayor's next lesson, then, is that he should not take the analyst's discount rate as given, because the rate expected now may not turn out to be the actual rate. An increase in discount rate necessarily decreases the project's NPV.

Take note, though, that the purpose of the bridge project is not to have a high NPV, just a positive one. At a toll of one dollar per crossing, assuming the lower discount rate, we achieve that positive NPV. And if the travelers are by and large willing to pay the one dollar, then the project is worthwhile (on internal grounds only—we have not yet gotten to external effects). After all, the purpose is to provide bridge services at reasonable cost, not to maximize toll receipts, which the city could do anyway just by installing toll booths at street intersections.

For decades now, the NPV technique for cost-benefit analysis has been a standard method by which to analyze an infrastructure project and, even more importantly, compare proposed projects. Several US presidential administrations have maintained an executive order requiring federal agencies to perform the analysis for federally sponsored infrastructure projects. The method does provide a logical and convenient framework by which to compare proposed projects. Yet the method is in no way a substitute for good judgment. Each cost-benefit analysis is only as good as the data and underlying assumptions on which it is based.

SENSITIVITY ANALYSIS AND RISK

The method we have discussed has a powerful additional application: it allows us to look at NPVs under alternative assumptions. We have already illustrated the project's sensitivity to an increase in discount rate from 3.5 percent to 5 percent: the net present value drops from a positive $1 million to negative $4 million. How sensitive is our bridge to other changes in assumptions?

Consider a scenario in which the digging of the ramps on one shore reveals a previously unknown Indian village site. This launches the state's procedures for reviewing archeological sites. It also gives legal ammunition to a group that wants to reduce development on riverfront land. Its lawsuit holds up construction for three years (during which there is a $1 million yearly expenditure to maintain the site) until a negotiated resolution that forces the city to move the ramps at an additional cost of $5 million. Under this scenario (table 7.3), with the discount rate remaining at 3.5 percent,

Table 7.3. Net Present Value ($ millions) of New Bridge
Under Alternative Scenarios and Discount Rates

Scenario	3.5%	5.0%
1. Base Scenario (Table 7.2)	1.0	−4.3
2. Construction extended by 3 years; project cost up by $8 million	−7.9	−12.9
3. Paid crossings per year decrease by 10%	−3.1	−7.6
4. Bridge service life is only 20 years	−7.9	−12.9
5. Toll increases by 50% to $1.50	21.8	12.3

the NPV drops to −$7.9 million. The bridge NPV is, in short, quite sensitive to construction delays and increases in construction cost.

Now imagine a scenario in which a downtown employer, housed in a skyscraper on one side of the bridge, goes bankrupt and closes up shop. At least 10 percent of drivers each year cross just to get to and from this skyscraper, or to and from other directly dependent businesses (restaurants, insurance companies, maintenance firms, etc.). If paid crossings decrease by 10 percent below the projected amount in the base scenario, toll revenues amount to just $2.25 million per year and, so under a 3.5 percent discount rate, NPV drops to −$3.1 million.

In still another scenario, the use of private cars declines drastically in the future, say at the twentieth year of bridge service life. By then, a long-sought new subway system will be built along with a future crossing, whether bridge or tunnel, making the trip across the river more convenient for many who would otherwise take cars or buses. What's more, gasoline and its substitutes will be shockingly expensive, drastically reducing vehicle traffic. In view of the vagaries of projecting more than 20 years into the future, scenario 4 assumes that the bridge has only a 20 years' service lifetime. The NPV at 3.5 percent interest now falls to −$7.9 million, clearly that of a bad project.

But a final scenario changes the whole picture. At one dollar a crossing, the toll is inexpensive. Even a 25-cent increase in the toll, though a small amount for each vehicle, adds up over all the crossings each day, 365 days a year. Let's hold out for even more, a 50-cent increase. At the 3.5 percent discount rate, a toll of $1.50 yields an NPV of $21.8 million. This bridge NPV is, in short, highly sensitive to toll increases. For proponents of the project, the benefits from even modest toll increases easily outweigh the costs from other, bad scenarios, such as construction delays.

In light of the uncertainties in making long-term projections, some analysts suggest that risk be explicitly incorporated into the analysis. So, for example, we should study the percentage of bridge construction projects in the past few years that have experienced delays and cost overruns. We should then adjust cost and benefit projections by the probabilities (garnered from our study) that bridge construction remains on budget. Though the idea of being explicit about risk is attractive, it does require added analytical time, and what's more, added assumptions about what inferences can be drawn from limited data. The past-project performances we observe in the study may in fact represent experiences during a different part of the business cycle, a different period in the history of car transportation and gasoline prices, in projects dependent on different construction equipment, across places that vary in the ways they manage projects and have variances in timeliness of construction and in cost overruns, such that the national probabilities of cost overruns in past years may not be representative of the local situation in a future year. As for our scenario on decreasing car use in 20 years, factual sources are nonexistent, though opinions, including expert opinions, abound. If risk analyses are incorporated into cost-benefit analyses, the decider's most important role is to ask hard questions about where the risk data came from.

A ONE-HUNDRED-YEAR BRIDGE?

Roman aqueducts are still sturdy after 2000 years, so one may well ask, why should our new bridge be scheduled to last for just 30 years? To be sure, our having set the service life at 30 years doesn't mean that the bridge will crumble the following year. It might be in good shape for much longer, even if periodic rehabilitations become increasingly expensive. That having been said, distressingly large proportions of modern bridges become structurally deficient in just decades, to the point that, in 40 years or thereabout, replacement makes more sense than further rehabilitation. They are just not built to last.

Shouldn't it be worth paying more now for a bridge that lasts, not a millennium (a duration we hope for later in the book), but just a century? Engineers are entirely capable of designing a bridge with materials that make it much more durable. Let's now analyze the costs and benefits of a long-lasting bridge. Project price (construction plus all studies, permits, inspections, etc.) jumps, let's say, by two thirds, from $27 million to $45 million; and on-shore road and ramp work by half from $6 million to $9 million. Construction still takes three years. Recurring costs and revenues are still set at constant dollars, with tolls assumed to stay the same as in the basic scenario. If we take equal care with preventive maintenance and periodic rehabilitation, the bridge should last over a century. Observe carefully: we've paid just 63 percent more for the bridge, but it will last more than three times as long. That should be a great deal. To check, let's examine table 7.4, which calculates the NPV for the 100 year bridge.

First note (see the net cash flow column) that, in years when no rehabilitation occurs, the receipts steadily outweigh operations and maintenance costs by $2.1 million. This bridge income remains steady over 100 years of service. Now follow the column for the 3.5 percent discount rate starting from the top. As the years pass, the annual net cash flow of $2.1 million is worth less and less in present-value terms. By the hundredth year of service life (103 years from the project start), the present value of the cash flow is

Table 7.4. Costs and Benefits of Long-Lasting New Bridge in Constant $1000

Year	Costs	Benefits	Net Cash Flow	Present Value @ 3.5%	Present Value @ 5.0%
1	−18,000	0	−18,000	−17,391	−17,142
2	−18,000	0	−18,000	−16,803	−16,326
3	−18,000	0	−18,000	−16,234	−15,549
4	−400	2,500	2,100	1,830	1,727
5	−400	2,500	2,100	1,768	1,645
Years 6–95 not shown.......					
96	−400	2,500	2,100	77	19
97	−400	2,500	2,100	74	18
98	−400	2,500	2,100	72	17
99	−3,400	2,500	−900	−29	−7
100	−400	2,500	2,100	67	15
101	−400	2,500	2,100	65	15
102	−400	2,500	2,100	62	14
103	−400	2,500	2,100	60	13
	−118,000	250,000	132,000	**−3,147**	**−16,239**

100 YEAR SERVICE LIFE

Discount Rate at 3.5% or 5%

PV=Present Value

NPV: Appears in two rightmost cells in bottom row

Note 1: Bridge project cost is up 66.6% (over the shorter-lasting bridge) from $27 to $45 million. Road and ramp work up from $6 to $9 million. Project cost of $54 million is distributed over a three year construction period.

Note 2: Rehabilitation work every 12 years until 99th year.

Note 3: NPV is shown in bold in two bottom right cells.

only $60. The NPV is in the negative by more than $3 million. The NPV is even worse at the 5 percent discount rate. By our discounting methodology, it is *not* worthwhile to pay an extra 63 percent to build a bridge lasting three times as long.

The Romans may have been great at stone construction and gladiator sports but were evidently handicapped in the noble art of cost-benefit analysis. Many analysts do hold that it makes little sense to extend the analysis far into the future, because traffic, maintenance cost, interest rate, and even choice of transportation mode (car, bus, walking, cycling, mopeds, street cars, and whatever else comes along) become increasingly dubious the farther we project. Yes, technologies change fast, travel patterns change, and business conditions fluctuate. In private business, investors would be rightly skeptical about tying up their money for a hundred years.

But should the same wariness extend to civil infrastructure, and to bridges in particular? Cities will still exist in a hundred years, even a thousand years we hope, and if they do, mobility will continue to be a human essential, even if it happens with electric sedans or hydrogen streetcars, and pedestrians will still want to get to the river's other side, so a graceful bridge will still be appreciated, maybe more than it was when still young. By discounting the far future through the cost-benefit method, and letting that analysis drive the decision, we may well be investing in planned obsolescence. We may be turning civil infrastructure into a species of disposable goods—garage-sale bridges for future generations.

EXTERNAL COSTS AND BENEFITS

Our study of bridge effects has dealt so far only with internal costs and benefits; these consist of costs to the bridge owner and benefits to its users. These internal effects are sometimes also known as *direct effects* or *primary effects*. Complex as it has been, this enclosed realm of internal effects is intellectually safe. Outside it lies a conceptual jungle, to which this book can be only a rudimentary guide. This world of external effects equally goes under the name *social effects*. These effects also have divisions with an even more complex nomenclature that includes *indirect economic effects* (also known as *economic impacts*, they're divided into secondary impacts, tertiary impacts, etc.), *neighborhood effects, economic development effects, environmental effects*, and *intangible effects*, which have overlapping and often ambiguous meanings.

Let's consider indirect economic impacts. Recall that the costs the city bears and the benefits users enjoy each count as the direct (internal) impacts. Because it has borrowed for the bridge, the city may be told that it cannot borrow more without a change in its rating; our bridge will then impose additional indirect costs on other projects the city might want to undertake. Once the bridge is built, the new traffic patterns have chain reactions along the road network, speeding up traffic near the downtown, reducing gridlock miles away and lessening travel time from the city to suburbs. These are indirect positive impacts. More city residents frequent the suburban mall while others move to outlying suburban houses, imposing negative impact on city business and housing. As users can now get around in less time, some choose additional paid work in the city, positively impacting their wages, and eat more while downtown, positively affecting restaurants. Trucks more easily supply city businesses and carry products from city manufacturers to the rest of the world, increasing business revenues.

These secondary effects in turn generate tertiary waves of cost and benefit, to those who build the additional house in the suburb or to the wholesaler selling additional food to the downtown restaurant. To ever more minuscule extents, these waves spread out, to the nation as a whole. Researchers who model such effects use giant interindustry charts in which changes in one sector, say trucking, are examined for their impacts on all other sectors, whether housing construction or restaurants, according to a procedure called input-output analysis. A detailed economic impact analysis would try to capture these effects. Some believe that studies with complete national and regional input-output charts can give the fullest answer to the question of bridge economic impact.

That being said, there are critics who question whether economic impact studies truly capture the full range of infrastructure effects. One set of critics focuses on neighborhood effects, here meant literally as the neighborhoods near the foot of the bridge. Residents near the landings may resent the added traffic that the bridge sends though their streets. The traffic causes extra noise, endangers the children walking to school, and adds to local air pollution, which some believe increases asthma rates. If the area under the bridge becomes derelict, it may also become a meeting place for vagrants, increasing crime, or just residents' fear of crime. Then again, the formerly abandoned railway yard (let's say that's what was there before the bridge) is turned into a fine landscaped park accessible to the public. Traffic is elevated overhead, removing some exhaust from the residents. Drivers who would formerly zip through the area now stop to patronize local businesses, increasing employment and incomes. The critics point out that it is difficult to accurately measure these varied effects.

Other critics points out that the market economy is dynamic and flexible. Recessions and growth spurts occur, businesses expand or contract, new products lines open up using new technologies, and businesses are started or closed. Workers shift to new work sites, suppliers to new buyers, truckers to new customers, tourists to new destinations. Mobility is essential to all these shifts, most of which cannot be predicted, since investors themselves have a hard time predicting what they will do next year. It follows, the critics say, that this kind of dynamic economy depends heavily on the existence of a high-quality transportation network, one that provides lots of choices for users. The usual cost-benefit analytical procedure, which assumes fixed present resources, will tend to *underinvest* in infrastructure, because it will tend to underappreciate the range of alternative possibilities to which future investment will respond and underestimate the degree to which future economic development (to which the bridge contributes) will expand the resources available. These critics would say that, with the bridge in existence, the city will enjoy an increased flow of resources by which to pay for bridges and other future projects.

Another group of critics argues, by contrast, that road transportation poses environmental harms that ordinary cost-benefit analysis leaves out. The former railroad land on which our bridge is built could have been put to other use; in our cost-benefit analysis we attributed to it no cost because the city owned it, but perhaps we should have valued it, since it could have been put to use as a wildlife preserve. What's more, building a new bridge increases dependence on automobiles rather than public transit. As the bridge initially reduces travel times, people can move farther from each other, as can businesses from customers and suppliers, imposing greater distances on travelers, forcing still more people to rely on cars, and these added cars soon recongest the downtown streets that were briefly freed up. With the additional cycle of congestion, more land has to be given over for parking, more pollution enters local lungs, the country depends more on fossil fuels, and more carbon flows into the atmosphere.

Roads with high-speed traffic also act as barriers to pedestrian movement (though bridges are different in this respect since they eliminate a barrier, namely that posed by the river, and increase pedestrian access as long as there is a lane for crossing by foot). More car dependence even causes more obesity because of lack of exercise. The critics hold that conventional cost-benefit analyses systematically omit these social costs, so they tend to *overinvest* in new road infrastructure.

Then there are intangible effects, including intangible social benefits. A beautiful bridge can become a city's symbol, an object of pride, for generations. It can enhance rather than detract from the neighborhoods at which it lands. If its structure is dramatic, and if it effectively frames the urban downtown or a nearby mountain, it thrills people traversing it. To those walking nearby or seeing it from a distance, it becomes a reason to stop and gaze. The bridge becomes part of the lived experience of the environment. When well designed, the bridge intangibly improves lives, to an extent that technical methods like cost-benefit analysis are not good at measuring.

HOW TO JUDGE A BRIDGE PROPOSAL

Capital investment in big public things like infrastructure projects is bound to have many effects. Those who want to understand the project's value have no choice but to make their way through this conceptual thicket. We have suggestions on how to make it through to the other side (to the decision), but the help we can offer is only partial. Research into project analysis and especially into external effects is ongoing, and the debates are not going to disappear soon.

For decision making about our bridge, the essential starting point is the internal cost-benefit analysis, calculated repeatedly to assess sensitivities to alternative assumptions. Whatever the additional studies that may eventually be done, this is the first and foremost step and it has to be solid. Much care must go into cost studies and projections of avoided travel time. If there is not much trust in data on vehicle costs, driver costs, and accident rates, then cost per avoided travel hour is a good basic indicator, though a full internal NPV study is better.

External factors should indeed be taken into consideration; some will make the project look better while others will make it look worse. So the finding that a project's internal costs and benefits had a positive NPV makes the project a contender for investment, but does not clinch the matter. Studies of external effects may yet show it to be a bad idea. Conversely, projects with negative internal NPV under multiple sensitivity studies should be assessed skeptically on whether further study is even called for. A road bridge that does not pass the internal cost-benefit test for its basic function as a traffic carrier will in all likelihood not be rescued by studies of external effects.

Neighborhood effects, economic development effects, environmental effects, and intangible effects are legitimate additional factors in decision. But they are typically hard to measure; different analysts arrive at different answers. If the bridge causes nuisances at neighborhoods near the landings, that's a cause for concern, but is not decisive. For those who reside in a city, fluctuations in traffic, air quality, crime, etc. are common—they may occur just because a new office building went up and shifted traffic patterns. For neighborhood effects to be decisive, the external effect the project imposes should be substantial, beyond some threshold we can't estimate here. Project supporters (and opponents) who claim neighborhood effect as their rationale should be able to show that the added benefit (or harm) is indeed substantial.

Neither is it decisive for the project decision to say that is has negative environmental effects. That the bridge will cause added air pollution is no fatal blow, because pollution might increase even more if the project is not built. This could occur in a place with population growth and an expanding road network. With the bridge unbuilt, the traffic distributes itself differently (than it would have with the bridge built) and leaves many cars idling in traffic jams downtown where the bridge should have been. To be sure, the time will come when the US highway system will stop expanding and when new modes of transportation will come to the fore, but that will be a momentous change in society, and present speculation about that future may not do much good now for a local bridge decision. Those who oppose the bridge on environmental grounds should show not just that cars cause pollution in general but that this particular project poses substantial environmental harms.

In one respect, though, the concerns about neighborhood, economic development, environment, and intangible effects coincide. Durable civic infrastructure makes for a predictable urban environment and stable neighborhoods. Ample long-lasting infrastructure provides a solid base for both a vibrant economy and a healthier environment. We strongly suspect that a bridge's greatest environmental disruption comes from being built and demolished too often—when the bridge has turned into a disposable consumer good, demolished and rebuilt periodically at great expense. And it is the durable and magnificent bridge that brings the intangible benefits: a lasting legacy and visible identity to a place. Cost-benefit analysis cannot capture this intangible effect. Considering a proposed bridge, deciders should rigorously inspect its internal cost-benefit, but should, thereupon, also reflect on the bridge as an enduring monument.

Further Reading

Richard Hudson Clough, Glenn A. Sears, S. Keoki Sears, *Construction Project Management*, 4th ed. (Hoboken, NJ: John Wiley, 2000) uses bridge construction as an extended example for explaining the complexities of cost estimating. A number of texts on capital investment introduce cost-benefit—for example Diana Fuguitt and Shanton J. Wilcox's *Cost Benefit Analysis for Public Sector Decision Makers* (Westport, CT: Greenwood, 1999). Anthony Edward Boardman and David H. Greenberg's "Discounting and the Social Discount Rate" examines some of the technical debates on the proper choice of discount rate for public projects, but a basic background in economics is needed; the article is on pp. 269–210 of Fred Thompson and Mark T. Green's *Handbook of Public Finance* (CRC Press, 1998).

Alvin S. Goodman and Makarand Hastak introduce discounted-cost benefit analysis alongside other techniques meant specifically for public works investment—see chapters 8–10 of their *Infrastructure Planning Handbook* (New York: ASCE Press and McGraw-Hill, 2006). For a simple introductory guide to cost-benefit analysis for road projects, along with a cost model for bridges, go to the California Department of Transportation (www. dot.ca.gov) and look for resources under its Office of Transportation Economics. The Federal Highway Administration's Office of Asset Management also has a short general book, *Economic Analysis Primer*, published in August 2003 and available for free download at the www.fhwa.dot.gov site. The Office of Management and Budget, Executive Office of the President, has on its webpage (www.whitehouse.gov/omb) the updated version of Circular A-94, originally issued in 1992, which has official federal guidelines for conducting cost-benefit analysis. More generally, the kind of analysis we have introduced here falls under the field of urban economics, on which a good text is Arthur O'Sullivan, *Urban Economics* (McGraw-Hill, in various editions, including 2011).

EIGHT

TRAFFIC ACROSS THE BRIDGE

The most important practical function of a bridge is to serve as a link connecting transportation routes across an obstacle, such as a river or limited-access highway, or to elevate the road so it does not interfere with activity below, such as a local street or natural ecosystem. For decisions on whether a bridge should be built, expanded, or rehabilitated, its effect on transportation is usually the foremost consideration.

For these decisions, the operative question is: What effect will the bridge have on improving transportation, say by decreasing travel times and reducing congestion? As we examine ways of answering the question in this chapter, we will look mainly at road transportation with cars and trucks, because that is the preponderant way in which people and goods move in the United States. We do care about subways, railways, trolleys, and pedestrian travel. We understand the advantages that transportation modes other than the automobile bring for better air quality, safer commuting, reduced use of fossil fuels, and streets made more pleasant for pedestrian enjoyment.

Still we will speak here of cars and related motor vehicles, because cars are not going away soon. For those who can afford to drive and are capable of driving, in places where road capacity is available, the car maximizes personal options for choosing destinations and choosing the time and route of travel. This flexibility is all the more important for businesses. In principle, if all flowed freely, and everyone could afford a vehicle, and there were no side effects like pollution, the car would maximize freedom, business and work opportunity, and personal satisfaction. Though for many urban planners, the first impulse is to want to reduce automobile dependence, we should not forget the value, even pleasure, that cars provide to millions of people.

BRIDGES AS LINKS IN NETWORKS

The freedom of movement that cars once seemed to promise could be fully achieved only if an extremely large proportion of the earth's land surface were devoted to vehicle conduits (roads and bridges), vehicle storage (parking), junctions and connections between conduits (intersections, ramps), and leaps across intervening conduits (overpasses and underpasses). In reality, there is not enough space in the world both for the ordinary requirements of human life and universal car accessibility.

Transportation routes are obviously constrained by the need for room for houses, backyards, agriculture, and so forth. They are constrained by landscape features such as lakes and mountains, and also by roads themselves, which interfere with the geometries of other proposed roads. In short, road proposals must be chosen under severe constraint of the existing land uses, built forms, landscape features, and the location of other roads. Over time, under various economic needs and political pressures, while working around these physical constraints, localities make often disjointed investments in road segments. The sum of these discrete decisions, whether made well or badly, results in a local road network.

Within the network, each road segment serves as a *link* between intersections and between various origins and destinations such as homes, office buildings, schools, and stadiums. Viewed merely from the point of view of efficient circulation of vehicles, it doesn't matter much whether the link lies directly on the earth's surface or is elevated. What matters is how efficiently the link serves within the traffic-carrying network.

So, from a traffic planner's standpoint, a bridge project must be evaluated much as any other road project, including a road-widening project. The operative question is the effect on traffic. Then again, even when it is viewed solely as a transportation link, the bridge is special. First, it is likely to be much more expensive than an equivalent stretch of surface road. Second, it is often a chokepoint for traffic.

Given these special features in road networks, proposed bridge projects must be evaluated rather carefully. Planners make that evaluation through the use of any of a number of transportation models. But before we get to that, let us first elaborate on bridges as traffic problems.

BRIDGES AS CHOKEPOINTS

That bridges serve as chokepoints (bottlenecks) in road networks is readily demonstrated. We have done so to our satisfaction by using the rankings of the private firm Inrix, which keeps track of real-time traffic congestion in the United States. The company periodically issues a scorecard for the worst traffic bottlenecks in the United States.

The company identified the worst 50 after analyzing traffic flow on over 30,000 road segments. Since we were in no mood to drive to all 50, we viewed them on satellite images such as those provided on major search engines. For each of the 50 road segments we asked: Does it incorporate a bridge component, including ramp, overpass or underpass (an underpass implying the presence of a bridge above it)? Is it directly adjacent to a major bridge? Of these 50 highly congested road segments, 41 had a bridge component or were next to a major bridge. For six we found no bridge and for three we could not tell from the maps. Our method was exploratory and informal, raising possibilities for error; we did not measure spans to make sure each elevated structure would be formally defined as a bridge. Still we became convinced that our hunch was right, that bridges are more likely than ordinary road links to be traffic chokepoints.

An illustration is the stretch of highway that ranks the very worst on Inrix's 2010 list of worst traffic corridors (figure 8.1 upper and lower), the Bruckner/Cross-Bronx Expressway, running east-west across the Manhattan and Bronx boroughs of New York City. Along this 11.3-mile stretch, free-flow travel time, available perhaps only in the middle of the night, would be 13 minutes. During the worst peak travel hour during the evening rush hour, however, it takes 63 minutes to traverse this stretch—a 50-minute delay.

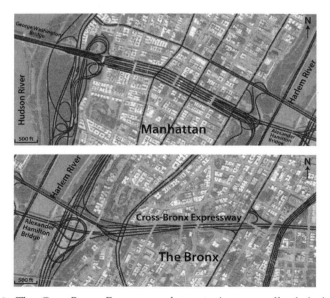

Figure 8.1. The Cross-Bronx Expressway, the nation's most traffic-choked highway, running east to west. In the upper image, it extends eastward from the George Washington Bridge, past complex ramp formations through Manhattan. As shown in the lower image, it continues across the Harlem River, past more ramps through the Bronx.

From an aerial view, some of the bottlenecks on the Cross-Bronx are readily apparent. At its western end, the Expressway intersects the George Washington Bridge and a north-south limited-access highway, and from their intersections sprout ganglia of ramps on which cars idle for exhausting minutes (figure 8.1). As it proceeds east, the expressway extrudes additional spaghetti clusters before and after it crosses the East River from Manhattan into the Bronx. Further east, the Expressway sprouts several other highway clusters connecting to still other highways, until one division of the highway reaches the Throggs Neck Bridge (not shown). Along its stretch the Cross-Bronx Expressway itself acts as a barrier to local north-south traffic, causing more localized congestion on streets alongside it.

Within a road system, why are bridges likely to be chokepoints? It is because of their cost, relative to the cost of ordinary roads. Where traffic has to cross a river or pass over a highway, the network density (the number of interconnected links per unit area) at either side of the barrier is likely to be greater than the network density crossing the barrier, because the cross-barrier link (the bridge) is more expensive than the regular on-land link (the stretch of road). In short, there is a relative scarcity of bridges as compared to other road links. Where many road links converge on the bridge, congestion builds up.

To sum up: as compared to equivalent lengths of road, bridges are special for transportation planning, not only because bridges are more expensive than roads but also because bridges are more likely to be chokepoints. More for bridges than for ordinary road links, decisions on where to build or expand must, therefore, be made with care. The decision must depend in part on the bridge's value in expediting traffic as compared to the bridge's expense.

THE NEED FOR MODELS

By US federal law in effect since the 1960s, transportation agencies that spend federal funds must consult transportation forecasts before spending the money. It stands to logic that it is important to do so. Before spending money, we should know where in the present network it is best to invest. Where is traffic slowed as from disrepair? To which neighborhoods is access difficult? Where does the most congestion occur? Where is safety most compromised? Where is the most air pollution caused? Where is the greatest increase in traffic expected ten years from now?

A transportation model should help answer such questions both by describing the present and forecasting the future—not an easy task. Our purpose in the rest of the chapter is to illustrate one model, by reference to a hypothetical place called Square City. But we must forewarn about the limits of models.

The transportation model should indeed help citizens and public officials decide what to do, but its purpose is not to automate the decision—to tell public officials what is the scientifically correct thing to do. Transportation models are not reliable or complete enough to provide such a solution. Rather, the model lets decision makers compare alternatives (say different alignments for roads) and rank alternative projects (which most reduces congestion, which costs least, which generates the least pollution). In a democratic society, the ultimate decision is political, and ideally should represent the judgment of properly elected representatives operating under the rule of law. Models are meant to inform their decisions and those of the persons they appoint.

Among the methods available for modeling traffic in an urban area, the most commonly used is the *four-step model*. It actually has two parts. The first part requires a description of the physical capacities of the area's road and transit network; estimates the numbers and locations of people who will use the system; and provides scenarios for the future of. The second part then estimates the patterns and volumes of traffic through which people travel from origins to destinations within the area. This second part of the modeling is done in four steps—for which this entire two-part model is named. We will now consider what it takes to construct these two parts of the model.

PART 1: CAPACITY AND USERS

Modeling the Present

The first part of our model should describe the present capacity the physical system (roads, intersections, subways, buses) has to carry vehicles and their riders in region; it should also tell us the numbers and locations of the system's users.

We should begin to describe the system's capacity with an inventory of the present road network. This may be as simple as a map that depicts major and minor roads and bridges and their interconnections. At a minimum, it must also include numbers of lanes in each direction and speed limits, but if we can afford further research, it should also have turn curvatures (that may slow traffic below its limit), areas with common slowdowns from accidents, efficiency of traffic flow at intersections, and other data. The combined information tells us the roads' capacity to carry traffic. Similarly, we should inventory the numbers of buses and subway cars, and their routes, average speeds and headways (time lapses between runs), and locations of stations, so we can estimate their capacity to carry passengers.

Now we want to know how many drivers and passengers use the system. Obviously, we need to find the size of the population and how it is distributed in the urban region. Less obviously, we also need to learn about the seemingly vague subject of "land uses." These are classifications of land into categories of predominant use, such as residential, retail, office, manufacturing, and institutional (schools, hospitals).

These categories are proxy indicators of where people are located during various times of day. To be sure, we could get a more accurate indication by conducting surveys again and again to figure out where people are. Geospatial technologies could eventually let us know everyone's precise location, if people don't mind being constantly monitored. In the meantime, land use maps are less expensive and less intrusive indicators of where people are located.

We know that people sleep and spend much of the weekend in areas in which the land use is residential. Land use data can also tell us which areas are more or less dense: which have high apartment buildings versus mansions with big lawns, and which have high-rise office buildings versus single-story retail strips. From land use data, including density data, we can estimate where—in which parts of an urban region—people sleep at nights, where they go in the daytime (among schools, offices, stores, courts, parks), where they go in the evening for entertainment (sports arenas, theaters, night clubs), and where trucks are likely to be destined (ports, factories, warehouses). Land uses are usually stable for years or decades, making them fairly durable indicators. Land use data is indeed useful.

Modeling the Future

So far we have been making a model that tells us only where transportation capacity and potential travelers are today. For a model that can help us make decisions whether to build or replace a bridge, we should also find out about capacity and travelers in the future, say fifteen years from now, because infrastructure projects take a long time to build and have service lives lasting decades. We particularly have to be concerned about population change and shifts of land use.

For population change, we can project into the future from the record of demographic growth or decline in the past few years—if the trend is up, the simplest assumption is that it will continue to go up. We should also be on the lookout for more specific demographic trends, such as the increasing preponderance of elderly people, or in-migration from other regions, or movements of people out of some neighborhoods into others, all affecting our projections of future travel volumes. Yet we know that none of these trends has to last; we are really not sure whether the trends will continue into the future. A current recession leads us to underestimate future growth; a war somewhere in the world causes an influx of new immigrants; and the opening of a new car-manufacturing plant escalates population growth far beyond our expectations. Alternatively, whole industries can die in a region, as steel industries did a few decades ago in parts of the US, causing huge out-migrations. These uncertainties are the very reasons we call this process "forecasting" rather than "prediction."

As model makers, we should also grapple with changing distributions of land uses. The growing immigrant neighborhood, the new food-distribution warehouse, and the office buildings and shopping malls all exert new demands on roads. We can try to anticipate by tracking new building permits, real estate prices (increased prices indicate more pressure to build or rehabilitate), and major real estate announcements, such as the decision to build a new airport or 50-story hotel complex. If we follow good forecasting practice, we will recognize the uncertainties and provide low, moderate, and high forecasts (say for population change) and let the decision makers decide which they most believe.

The Need for Scenarios

Not the least of the challenges when it comes to forecasting urban futures is a subtle paradox. The local officials, business people, and leading citizens who now want the forecast (say on the future growth in density of a residential neighborhood) are themselves in part responsible for helping bring about the outcome in question. That is, the people who want future land use information are in part now responsible for making the decisions that generate the future. For example, public officials may be responsible for changes in ordinances to permit denser housing; land developers may be ones to invest in the building of new housing; and activists may oppose growth that cuts down trees and increases local traffic.

This is why public participation in the process called *scenario building* has become so important to land use planning. Representatives and stakeholders from the region are asked to attend meetings and take part in exercises with urban planners to foresee the kind of city they would like to have. Ideally, their preferences get worked into regulations guiding future growth. The regulations may, for example, determine where tall office complexes or dense residential projects will be permitted. What is important in such scenario efforts is not to get random public guesses and opinions about what the future will hold (an aggregation of uninformed opinions won't do much good) but to get citizens' involvement in present decisions on the locations of future land uses. These present decisions lend a measure of stability to future trends and improve forecasting.

Then again, economic conditions change and opinions shift. The mayor and council members may now commit to a land use plan, but next year, when an investor offers to open up headquarters on the waterfront for a giant software firm, the legislators can quickly drop the old plan and substitute a new one. So it is that transportation modelers should make no claim of predictive precision—the transportation forecast should not be confused with the engineering predictions that tell us how a satellite will enter orbit. Rather, the forecasts help us more clearly assess the potential values of a project, such as a bridge, under different future scenarios. To assess the value more completely, we have to go on to the second part of the modeling process.

PART 2: TRAVEL PATTERNS—THE FOUR-STEP PROCESS

Having assembled a description of road and transit capacities, and of population and land use, and also scenarios for the future, we can go on to the rest of our effort: to figure out the patterns of traffic by which people get from their origins to destinations in our region. We will use the "four-step" process for which this entire modeling method is named.

To set up a four-step model, we must divide our metropolitan area into *transportation analysis zones*, known as TAZs. Here we just call them "zones," but specifically mean TAZs, not any other kind of zone. Depending on the complexity of the model, a zone may be as small as a few city blocks or as large as a rural township. Every part of the metro's land area should be placed in a zone, except maybe a wilderness area.

We set up these zones because we are not going to try to estimate individuals' travel patterns from their specific home addresses to their various specific destinations, whether school or theater or workplace. As of the time when we are writing (when methods are changing because of new geospatial technologies), we do not have such information and do not wish to snoop into peoples' lives. Instead, for the particular metro area we are working on, we estimate trips not for actual individuals but for average behavior by occupants of zones.

Before we estimate travel between our zones, we should remember that many trips come into or head out of the region: they have origins outside our metro area, or have destinations outside it, or are just passing through but using local infrastructure. To account for these, modelers also establish *external zones*. Typically, over 90 percent of all trips taken in a metro have origins and destinations within the metro—trips from and to *internal zones*.

In recent years, the National Capital Region Transportation Planning Board, which serves Washington, DC, has used a model that divides its metro area into 2200 zones and recognizes 28,000 road segments, plus many transit lines. The model we soon present is rather smaller than Washington's. It is the model for an invented place called Square City, which has only

12 zones, road segments in the dozens, and no transit at all. Square City is of interest to us because the City Mothers along with the City Fathers are wondering whether a new bridge is needed.

As almost all the world's cities are parts of complex metropolitan transportation networks, often with extensive suburbs, Square City is unusual, but conveniently so. It has barely any suburbs. Conveniently, it is also square, so it can be divided into four equal-sized internal zones and—to represent the origins and destinations in the rest of the world—eight external zones, as seen in figure 8.2. The river that is causing the bridge problem runs north to south through the center of the city, dividing the NE and SE zones on the east bank from the SW and NW zones on the west bank.

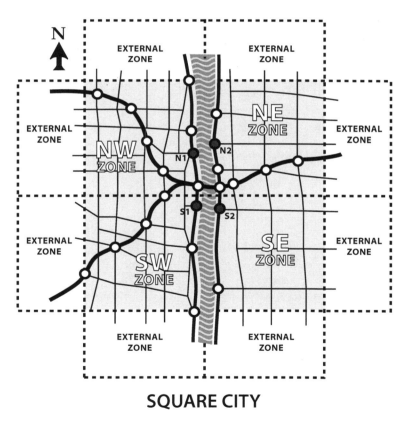

SQUARE CITY

Figure 8.2. Square City divided into Transportation Analysis Zones.

Two highways come in from the west and converge near the center of the city on a single highway crossing the river on the city's single existing bridge. Two additional highways line the river, one on each bank, intersecting the east-west highway. The combined population of the west bank zones is twice that of the east bank, but the City Parents are concerned that rapid growth on the east bank will change the patterns of commuting in the city. The most widely believed forecast is that population will grow rapidly in the SE zone.

After the zones are delineated, the modeler can begin the four-step process. First, in the *trip generation* step, she estimates numbers of trips each zone sends to other zones and the number it receives from other zones, but without yet specifying which these other zones are. In the second step, *trip distribution*, she estimates the zones to which each zone sends its travelers (and reciprocally the zones from which each zone receives travelers). In the third step, *modal split*, the modeler offers estimates on which percentage from each zone travels by which transportation mode, whether car, subway, or walking. In the fourth and final step, *trip assignment*, she successively fills the roads and subways with hypothetical travelers until conveyances are filled up, congestion occurs, and alternative routes have to be found.

If all has been done well, this final step yields a picture of the metro area's traffic pattern. It is then that the model can prove its value. By hypothetically widening a road, increasing car occupancy (reducing the number of cars), or adding a bridge, the modeler can provide forecasts of the effects of alternative policies on traffic. Let's now go through the steps one by one.

Trip Generation

Our first step is to figure out how many trips each zone sends to other zones and how many it receives from other zones. The trips sent out are known as the zone's *production* and those received as its *attraction*. Travel that has its origin and destination within the zone is excluded; this is simply a limitation of the four-step model.

For residential land uses within a zone, we estimate trip production from population or number of households, adjusted by average car ownership. For industrial and office areas in the zone, we estimate trip attraction, understanding that those attracted may include workers, sales and delivery people, and customers.

There are three ways to get this data. First, we could conduct an extensive survey of households in the city, enough to get accurate results for each zone. We would ask households how many trips they take and where they go (what their destinations are) for various activities (work, school, doctors visits) at different times of the day, and perhaps different times of the year. This would be the best approach were it not the most expensive, and if people would indeed have the patience to provide so many answers. In practice, this kind of direct data has been sparsely if at all available.

As our second option, we can consult data that researchers have collected for various parts of the country on average sizes of households, numbers of cars owned, and numbers of trips per day for job, shopping, and social visits. What we get is propensity to travel by various household characteristics. We learn that two-person households owning one car typically take 2.4 trips per day. We then find how many two-person single-car households there are in our zone of interest, and then multiply that number by 2.4; and then repeat the procedure for each other combination of household size and car ownership.

As our third option, we can consult trip-generation rates by land use, collected by underpaid students standing on streets and observing numbers of trips in and out of buildings. Such rates have been assembled for decades and may be obtained in printed and electronic formats, often called *trip generation manuals*. For a shopping mall with a given number of retail establishments or given square footage, the rates tell us the average numbers of trips in and out. To be sure, it is wise to verify national data through local trip-generation studies.

Now comes the more conceptually complicated part of the trip-generation step: reconciling trips produced with trips attracted (excluding trips that start and end inside a single zone). The number of trips produced by households in all zones during, say, a one-hour slot in the morning must equal the number of trips attracted by jobs, schools, shopping centers, and other households (some people may just be visiting friends) in all zones during that hour. Among the difficulties are trips that cut across the one-hour slot and chains of trips by one traveler within the hour. The need to reconcile the production and attraction numbers adds accuracy to this method.

For Square City, we have kept our model exceedingly simple: we assume that zones have identical proportions of residential, retail, and other land uses and that trip production and attraction are completely proportional to population.

Trip Distribution

Imagine we are going through our four-step model for a hypothetical metro region. In the morning peak rush hour, of all the trips Zone A produces (they leave Zone A for other zones), 50 percent go to work, 30 percent to school, 10 percent to shopping or professional services, and the rest go elsewhere for other reasons. If we had perfect data on the zone, we would also know the zones that are attracting its residents.

Without direct data, we must use some method to estimate the zones to which people go. Let us start with trips taken to work. By the most commonly used method, the modeler assumes that the work-travelers leaving Zone A distribute themselves among Zones B, C, D, etc., by those zones' attractiveness (how many work attractants they have) and their distance from Zone A (road distance, sometimes adjusted by tolls or transit fares). The farther the work from Zone A, the less likely that the Zone A residents will work there. Let's switch now to shopping trips: the farther the shopping site from Zone A, the less likely that Zone A residents will shop there. Known as the *friction factor*, the rate at which attractiveness declines over distance must be estimated to make the model work. (Because attractiveness is assumed to decline with distance, it is usually called a *gravity model*.)

We know that these are crude assumptions. Doctors and business executives in a secluded neighborhood of mansions may travel far from their suburban homes to hospitals and corporate offices, while blue-collar workers from an old urban district may travel far in the opposite direction to work in an industrial area, while young or poor service workers may find jobs closer to their homes. As matters of income and profession greatly affect destination choices, modelers have over the years tried to incorporate ever more factors, beyond attractiveness and distance, into the models, though it has not always been clear that multiplying the numbers of variables increases accuracy.

In Square City, everyone is of the same economic and social group, very much simplifying our model.

Modal Split

Now that we have estimated where the current inhabitants of a zone are going, we ask how they get there, whether in a single-occupancy car, car pool, or bus, or by rail, walking or bicycling—each of these ways called a transportation *mode*. As hopes have grown that Americans will quit their cars and instead walk or cycle, and solutions have been sought for congestion and petroleum reliance, ever newer ways have evolved for trying to estimate modal choice.

As a start, we need to find out what modes are at all available between one's zone of origin and the destination. If there is no subway, no bus, and no sidewalk, but everyone owns a car, then there is not much of a modal choice. The modeler will still want to know likely car occupancy, since that affects the volumes of vehicles that will load the highways. The easiest way to estimate is to consult survey data on car occupancy by income, by number of cars owned, or by type of destination.

Let us assume that there is indeed a subway station and a person prone to exhaustive decision making thinks about which mode to take for the trip. He considers the mode's cost, namely the subway fare, versus the amortized cost of his car plus gasoline plus insurance plus tolls, plus parking. He also considers travel time in the car versus the combined time for getting to the subway station, waiting for the ride, the ride's duration, and then getting to his destination. He may count traffic accident danger and inability to read while traveling as a disadvantage of the car; heat, crowding, and uncertain delays, as problems with the subway. Cultural preferences, like fear of motley crowds or dislike of carbon-spewing cars, may also come into play. Though it would be an unusually obsessive traveler who actually carried out such detailed calculation before taking a trip, large numbers of travelers do implicitly, in varied individual ways, take such matters into consideration. However, the modeler's attempts to include ever more such variables in the model, and estimate causal relations between them, would add ever more complexity without necessarily improving accuracy.

Square City has, however, earned its name. No one walks or cycles, there are no subways or buses, and no one has heard of climate change. All travelers drive single-occupancy cars.

Trip Assignment

Since we know from previous steps how many are traveling from which origin zone to which destination zone, using which transportation mode, we can assign them to the road and transit network we have already inventoried. Different models do the assigning in different ways, but they usually start with the assumption that travelers take the quickest route to their destination. Starting with one traveler and then going to a second, the model selects the quickest path for each. When a road segment gets filled up, the model estimates the slowdown in traffic. As traffic slows, the traveler may no longer be on the quickest path, so the model reassigns the person to an uncongested path. Similarly for transit: the traveler is assigned to a bus or subway car on which space is available, given numbers of vehicles and headway times. The process is repeated until trip volume gets close to transportation capacity.

Let us now run our traffic assignment model for Square City during the peak rush hour. To keep things simple, we assume that at the bottom of the hour the roads are empty. Within each zone, in proportion to zone population, our model loads the cars onto local streets, from which it channels them onto highways. As traffic slows below the speed limit, the model reassigns cars to alternate routes, simulating drivers' decisions to keep moving fast. If it cannot find a way to balance car numbers with road capacity, the model can be run through an additional iteration, in which it loads cars along different routes (as drivers themselves would try alternatives, based on experience) to pick less congested routes.

In our model of Square City, we continue our runs of the model until the tenth iteration. We then consult the model to tell us the speeds on specific road segments. We are particularly interested in the existing bridge, known as Old Bridge, which is Square City's chokepoint, where the speed limit is 45 mph. Figure 8.3 shows minute-by-minute average speeds on the bridge during rush hour. We see that traffic on the bridge starts off at the free-flow rate of 45 mph. As traffic builds up, average speed declines, though at different rates for each of the two traffic directions. After 25 minutes, the traffic is creeping at 8 mph, and remains at that speed until the end of the hour.

The software through which we run our model can also generate maps of traffic location in Square City. As we would expect, the maps (not shown) of conditions during rush hour show traffic backing up at approaches to the bridge.

SQUARE CITY UNDER FOUR SCENARIOS

We have just described the results under the current population and the single existing bridge. Let's call this scenario 1, the base scenario. Being

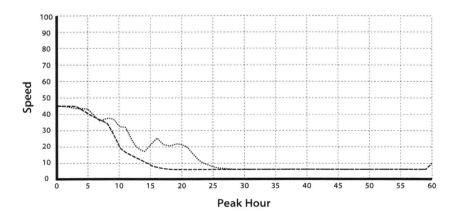

Figure 8.3. Average traffic speed (in each direction) on Old Bridge during peak travel hour.

careful modelers, we have checked the results against actual traffic speeds in Square City and found the model to be accurate. But Square citizens will rightly shrug at being told what they already know, that traffic is awfully slow at the bridge during rush hour. Rather, they want to find out what will happen under alternative scenarios.

We use our model to examine three additional scenarios. Under scenario 2, the city's population stays the same, but a new bridge, called North Bridge, has been built to the north of Old Bridge (connecting positions N1 and N2 in figure 8.2). It is positioned there because the soil conditions are good for bridge piers and because the city owns land at the proposed landings. Under this scenario, the average speed of rush hour traffic on Old Bridge has gone up to 44 percent of the speed limit (around 20 mph), and North Bridge operates at a healthy 85 percent of the speed limit.

Now scenario 3 examines a possibility that has local officials concerned. Observing the availability of open land and increasing land development in the SE zone, they suspect that population there will grow rapidly. In scenario 3, population has doubled in the SE zone, making it equal to the populations in each of the two west bank zones. Average speed on Old Bridge drops to even less than it was before the new construction—to just above 7 mph. Traffic on North Bridge is also faring worse, down to 67 percent of the speed limit.

The expected growth in the SE zone suggests that North Bridge is in the wrong position. So scenario 4 offers the same population increment as the previous scenario but replaces North Bridge with South Bridge (between S1 and S2 in figure 8.2). The results from the four scenarios can be seen in table 8.1.

Scenario 4 forecasts what happens if the city builds South Bridge. The result is counterintuitive. Average peak-hour traffic at South Bridge (which is positioned next to the fast-growing SE zone) moves at 67 percent of the speed limit, just as North Bridge traffic did in the previous scenario. One possible reason is that much of the traffic to and from the SE zone takes the east-bank riverside highway to Old Bridge for access to the center of the west bank, from which both highly populated western zones are accessible. What South Bridge does do is take the pressure off Old Bridge. Whereas in scenario 3 Old Bridge traffic averaged 15 percent of speed limit, in scenario 4 it performs better, at 35 percent of speed limit. The counterintuitive result illustrates the value of this kind of computer model: that it can reveal findings we might not anticipate just with superficial common sense.

These results provide valuable information: they tell Square City citizens what they would gain by alternative capital investments under alternative growth scenarios. Now they have to decide whether the traffic improvement brought by South Bridge is worth the increased construction

Table 8.1. Four bridge congestion scenarios for Square City during peak traffic hour

Scenario	1	2	3	4
Bridge	Old Bridge only	Old Bridge and North Bridge	Old Bridge and North Bridge	Old Bridge and South Bridge
Population by zone	NW: 40,000 SW: 40,000 NE: 20,000 **SE: 20,000**	NW: 40,000 SW: 40,000 NE: 20,000 **SE: 20,000**	NW: 40,000 SW: 40,000 NE: 20,000 **SE: 40,000**	NW: 40,000 SW: 40,000 NE: 20,000 **SE: 40,000**
Average congestion at Bridge—% of speed limit	Old Bridge: 20% [No other bridge]	Old Bridge: 44% North Bridge: 85%	Old Bridge: 15% North Bridge: 67%	Old Bridge: 35% South Bridge: 67%
Avg. Travel Time (mins.)	9.2	8.2	9.4	8.3

Note: Average congestion measured between 10th and 50th minute. Peak congestion measured at peak traffic minute. Model developed by Jinge Hu.

and land-acquisition costs for it as compared to North Bridge. They also have to debate whether the higher price of the South Bridge is worth the overall travel time improvement for the city—it is only a one-minute average improvement during rush hour over the North Bridge option. They could make a better decision if they consulted both this transportation model and the Net Present Value method introduced in the previous chapter.

TRAFFIC MODELS FOR INFRASTRUCTURE DECISION: SOME CAUTIONS

Experienced transportation planners will point out that, even if the model correctly depicts the present and alternative future scenarios in Square City, it is easy to go wrong with it. Perceptive readers will already have picked out some glaring problem. Maybe instead of building a new bridge, the city should invest in a transit system that relieves congestion and gives Square residents new options. The same goes for investment in walkways and bicycle routes. After planners go to the trouble of preparing a model, which inevitably requires some assumptions and simplifications, they and the model's users sometimes forget those assumptions and restrict their decisions only to options that the model makes available. The unwary policy makers may then be led to neglect alternatives that the model did not analyze.

Even if we disregard Square City for the moment and think instead about a more realistic place in which some people do walk and bike, and we build a four-step model that does include modal choice for nonmotorized transportation, the model will still incorporate a subtle assumption that discourages investment in pedestrianism. It is the assumption that transportation is meant to get people from origins to destinations as fast as possible. Applied to walking, this is a faulty assumption because in our time walkers often walk for pleasure, recreation, and health. Misdirected by the origin-destination model, the bridge engineer may make no pedestrian crossing or make just a narrow and unpleasant one, missing the opportunity to give citizens a pleasant experience of their city.

Now let us move to an even more encompassing caution: we have responded to traffic congestion by suggesting new infrastructure investment, a new bridge in one location or another. In short, we have assumed that the way to deal with congestion is by increasing infrastructure *capacity* (or supply), an expensive option that can actually, paradoxically, increase the density of cars on the road system.

Transportation planners have learned this lesson again and again. After repeated efforts in the United States to reduce traffic congestion by building a new ramp, adding a lane to a road, or building an additional bridge, they have repeatedly discovered that it reaches its capacity faster than the forecasting models predicted. This phenomenon (to which we also return in the next chapter) is sometimes called *induced demand* or *latent demand*. Potential motorists, it turns out, previously avoided car travel because of slow, annoying traffic. Once the new bridge is built and traffic starts to speeds up, the formerly latent motorists enter the roads, until the road gets recongested, finally disinclining further new motorists from the drive. If in Square City, too, the new bridge induces additional driving, then the city's two bridges might soon be as congested as just one bridge was previously.

Perhaps what we should do instead of building a new bridge is reduce or otherwise manage travel *volume* (demand for the infrastructure). Ways of reducing traffic through demand management abound. Tolls on the bridges can be raised during rush hour to discourage use during peak hours; employers can be encouraged to introduce flextime, so fewer people hit the road during those hours; city parking rates can be heightened or taxed to increase the incentive for taking transit; and high-speed lanes can be installed to encourage car-pooling.

Then again, Square City should be cautious about rushing to solve traffic problems with bicycle routes and sidewalks. Even under the phenomenon of induced demand, if the bridge generates more traffic than expected, the city's road system afterward does support more commerce (more comings and goings) than it did previously, and this may be good for the economy, even though movement across bridges is still slow. Also, the city now has few suburbs. Americans have long exhibited great reluctance to leave their cars, or even share them; car-pooling efforts usually flop. If city traffic and parking become much more difficult, some employers may move to the suburbs, resulting in even more complex and expensive needs to spread out the road infrastructure, adding further to auto reliance.

As we write, worries about global climate change and fossil fuels lead many to believe, or hope, that public transit will increasingly replace private cars. It is at least as likely that the car in America will remain supreme, whatever the energy source on which it runs. Warm feelings about walking and biking should not lead us to obstruct "commerce," meant in the old sense of the word, to encompass all comings and goings of society. Commerce builds economies, enriches cultures, and brings distant people together. As important as it is to have room for walking and biking, they can never be the basis for modern commerce.

Transportation needs impose hard decisions on communities. Complex and contentious debates ensue about what should be done. A transportation model does not solve the problem of what to do, but does serve as a tool by which, if it is intelligently used, decision making can be improved.

TRAFFIC MODELS: LIMITATIONS AND PROSPECTS

In offering some cautions about traffic models, we have so far kept to the pleasant notion that, within the limits of their assumptions, they are accurate. Let us complete our discussion by acknowledging that often they are not.

To estimate trip production, we have relied on indices of average trips per household per car or average trips per retail establishment of a certain size. These are crude assumptions. After all, trips per household are subject to great variance: some households of given size and number of cars drive much more than others. Sometimes the reason for variation is economic: in some dense neighborhoods, the high cost of maintaining a car or to take certain trips gives travelers the incentive to switch modes. In other places, the reason is cultural. In immigrant neighborhoods where family links are extensive and spread far and wide, or in neighborhoods catering to some age groups (young, alternative lifestyle places), households may make more or fewer trips than average, to destinations that the simple models do not correctly forecast.

One answer is to increase the number of variables in the model: add more economic and social factors. In the effort to do so, much research is being done and important advances have been made. But effort has also gone into creating ever more elaborate models, which weave together ever larger numbers of variables, with ever more speculative functional relationships (with what percentage likelihood does a lower middle-class household with members holding three jobs under recession conditions choose multi-occupancy versus single-occupancy trips to varied zonal destinations?) but imperceptible improvements in accuracy. The more that the models seek to incorporate the economic and social determinants of travel choice, under an ever changing economy and society, the more difficult the modeling becomes.

What is most reliable in a transportation model, we have found, is the physical relationship between road (or transit vehicle) capacity and trip volume. We know pretty well how many cars can fit on a road or how many passengers into a subway car. We know less well the social and economic determinants of travel demand. It is troubling to have to make a model ever more elaborate in estimating human behavior, when the economic and sociological disciplines are themselves in flux and full of internal disagreement. As transportation planners, we may just have to reduce expectations and accept that transportation models are tools for exploring alternatives, and far from being the outcomes of a science of travel behavior.

There are, nonetheless, important opportunities to make travel models better, and thereby to make better infrastructure decisions, including bridge decisions. The most important is the development of geospatial tools that can tell us where people or vehicles are at various times of the day. To be sure, there are great hurdles. Researchers must assure that peoples' privacy is protected and that they do not fall prey to malevolent observers monitoring their movements. But the technology does open up the possibility that transportation models will come to be calibrated less from simple assumptions about human behavior, and more from real-time data on trip volume, trip origins and destinations, and infrastructure capacity.

Further Reading

The Inrix company's rankings of road segments with terrible traffic congestion is found on its website, www.inrix.com. A widely used textbook on the details of the four-step process is Juan de Dios Ortúzar and Luis G. Willumsen, *Modeling Transport*, 4th edition (Hoboken, NJ: John Wiley, 2011). A good introduction to the general reader is Edward Beimborm, Rob Kennedy, and William Schaefer, *Inside the Black Box: Making Transportation Models Work for Livable Communities*, published by the Center for a Better Environment and the Environmental Defense Fund (n.d.) and available for download at various sites on the Internet. Still another convenient source, copyrighted and posted by Oregon State University, Portland State University, and the University of Idaho, is the *Transportation Engineering Online Lab Manual*, which can be found by searching under that title; see its section on Traffic Demand Forecasting. The National Capital Transportation Planning Board's magazine *Region* introduced Washington's use of transportation modeling in its 2003 issue (vol. 42), pp. 19–25. For examples of a *Trip Generation Manual*, see the volumes by that name produced by the Institute of Transportation Engineers in Washington, DC.

NINE

THE BRIDGE IN THE ENVIRONMENT

Though bridges are large and obvious artificial objects in the environment, their consequences for the natural environment's health are usually not extreme. As compared to buildings, they are not heated or air conditioned, do not circulate piped water for occupants, and are not heavy users of energy, except for lighting. Their most obvious effect on the environment, the exhaust from the traffic that crosses them, is indiscernible from that of ordinary road traffic.

Most bridges in America are simply elevated highway spans over dry land or are highway ramps. When a new one is proposed at the site where a previous one is to be demolished, the environmental effect arises simply from increased traffic, from the energy consumed in making the steel or concrete, and from the construction process, with few if any additional deleterious effects on the environment.

Where bridges are meant to cross water bodies or wetlands, however, environmental consequences must be examined with care. And so they are in the United States (and in many other countries), by the force of federal and state laws: bridge projects at new sites must undergo a series of studies and public reviews to assess environmental impact. Central to this review process is the issuance of an *environmental impact statement*, abbreviated EIS.

This brings us to an ambiguous feature of the EIS. It is, on the one hand, a legal requirement, subject to complex bureaucratic procedures. On the other hand, it is in its basic structure a well-ordered logic for reviewing environmental effects. For anyone considering a significant infrastructure project, even one that is hypothetically operating outside this legal framework (say entirely on private property, with no effects on neighbors), the EIS logic provides a good way to judge the project's environmental effects and whether enough is being planned for alleviating harms.

In this chapter we briefly introduce some of the complex regulatory requirements that have accrued over time in the EIS process. More so, we describe the underlying logic that anyone should pursue when assessing a proposed new bridge crossing a water body.

It is widely recognized that the EIS process is better at identifying harm and possible abatement of such harm than at giving incentive for projects that positively benefit the environment. So later in the chapter we examine ways in which bridge projects could also take us toward that widely discussed but seldom (or inconsistently) defined ideal of sustainability. In particular, we will state our ideal, not the mere hundred-year bridge that we mentioned in chapter 7, but the thousand-year bridge.

GREAT LAKE CITY WANTS TO BUILD A BRIDGE

Let us consider an imaginary city on the US side of one of the Great Lakes—let us call it Great Lake City—at a once-thriving harbor. A river meanders roughly west to east, widening into the harbor just before it flows into the lake. On the south side of the harbor lies the city's downtown. On the north, acting as a barrier from lake waves, is a promontory that once housed steel mills, factories, and harbor facilities, and still has boat yards. A stretch of the promontory's south shore is a wetland. As the old industrial users have disappeared, the parts of the promontory that are not vacant have turned into a natural reserve and bird sanctuary, and the rest into areas for recreation, such as biking, wind surfing, and bird watching. Much of the area remains vacant. There is investor interest in developing some of the vacant lands for waterfront housing.

The one-time industrial users of the promontory had access to it through lake freighters and with rail lines and highway routes making the roundabout connection some distance to the west from the city's downtown. For recreational users and potential promontory residents, the old road links are too inconvenient. What is more, Lake City wants to increase residential, recreational, and business activity around its downtown harbor overall. Many urban planners and public figures hold that a new bridge will stimulate development around the harbor, helping revitalize the downtown. The bridge would also open up the opportunity for connecting scenic lakeside walkways and bicycle routes around the harbor (figure 9.1).

Great Lake City's consultants have identified three alignments along which the bridge could connect the promontory to the rest of the city. The preferred one would span a neck in the harbor, to the city's downtown. Since the city will need state and federal funds to build the bridge, and the bridge alignment extends a state road, and it crosses a waterway regulated

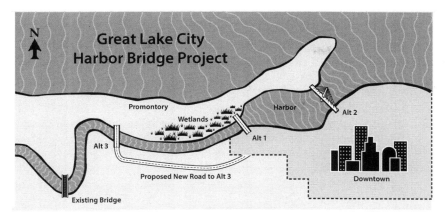

Figure 9.1. Bridge Proposals for "Harbor Bridge" in Great Lake City.

under several state and federal statutes, and for other legal reasons, they must prepare and submit an EIS.

PREPARING THE EIS

In speaking of the "city" doing so, we are simplifying. In reality, many agencies would be involved in the effort right from the start: the State Department of Transportation, federal highway officials, a local transportation coordinating body, a public development authority, environmental and maritime agencies, as well as the city and county governments. But one must be designated the "lead agency." So to make the story simple, let us say that it is the Great Lake City government itself that will lead the preparation of the EIS.

The city must, as one of its early steps, publish a notice that formally announces its intent to prepare an EIS and provide contact information. Environmental advocates, freight ship operators, land developers, public works agencies, the Coast Guard station patrolling this part of the Great Lakes, boaters groups, and downtown business groups, among many others, are thereby put on notice. Collectively, they are known as *stakeholders*. As an early step, the city will also have to start what is known as *scoping:* holding meetings and issuing documents through which the city and the stakeholders figure out what environmental issues need to be considered and what studies commissioned.

Based on the scoping, the lead agency works with other agencies (and stakeholders) to outline research tasks, allocate tasks among agencies, and commission studies. Often these are called *cooperating agencies,* on the hope, often well founded, that they will be indeed cooperative. We will learn more about this complex process in the next chapter, in our discussion of bridge delivery. In the meantime, let us say that after a year of study the lead agency submits the document officially termed the *Draft EIS.* For large or complex projects, this is a correspondingly large document, made available to the public in agency offices for inspection and often posted on the Internet. Hearings are then held and written comments received, not to mention criticisms. In response, the lead agency makes amendments or commissions further studies.

Results are then put together into what is officially called the *Final EIS,* though that term comes to seem premature if lawsuits, newly discovered environmental problems, or political pressures force the agency into amending it. From the initial notice through completion of the Final EIS and formal ruling to accept it, the process can take several years.

In the United States, a long history of legal rulings has expanded the coverage of the EIS, so that it isn't just examining effects pointed out by experts; it is also reviewing the project's fit with state and federal laws. Let us say that our Great Lakes City is in New York State. An EIS process for our proposed bridge would have to take into consideration some of the following federal and state laws or administrative regulations:

Federal Clean Water Act
New York State (NYS) water quality regulations
Federal Rivers and Harbors Act
Federal and state regulations governing wetlands
NYS designated environmental "areas of concern"
US Coast Guard permitting for navigable waterways
Federal Emergency Management Agency regulations for
    floodplains
NYS and Federal Coastal Zone Management Programs
NYS regulations for protection of aquifers and drinking water
    supplies
State and federal laws controlling spillage of untreated storm
    water (as over the new bridge into the harbor)
Laws protecting wildlife and fisheries
Laws covering threatened or endangered species
NYS Department of Transportation procedures for handling
    any hazardous materials
Federal Highway Administration noise abatement criteria

NYS Department of Transportation analysis to assess effects on
   energy consumption
Federal Clean Air Act standards
Administrative rulings for state or federally designated parks,
   recreational areas, heritage areas, and natural landmarks

The above is a partial list. It illustrates, of course, the bureaucratic
complexity of EIS. Despite the good intentions of the laws, the legal wel-
ter poses a potential problem for environmental policy. Whereas laws may
require painstaking examination of some problems, say lightly polluted rain
and snow running off from new pavements into the harbor, the same laws
may not be specific or detailed on the effects of, say, disruption of heavily
polluting toxic sediments buried under the river silt. So the force of the
rules may cause disproportionate attention to some environmental effects
as compared to others.

The EIS process has also expanded to include more than strictly
"environmental" issues, in the usual sense of the word. The EIS must now
consider the effects of the bridge on nearby urban land use (will it disrupt
a residential area? divert traffic from a commercial street?), architectural
landmarks, "neighborhood and community cohesion," social groups such
as minority and low-income populations and the disabled, school districts,
places of worship, regional and local economies, and business districts.

It is straightforward enough to identify landmark buildings on registers
of historic places. But effects on "neighborhood and community" are notori-
ously difficult to judge and are subject to multiple viewpoints about what
study method is appropriate and what impact is worthy of concern. Studies
of effects on "regional and local economies" can range from the simplistic
to the extremely complicated, with results varying with data sources used
and the theories of the economists hired.

To be sure, these expansions of EIS law have admirable intentions.
They mean to ensure that infrastructure projects take into account community concerns that are not strictly environmental, say by leading planners
to make that bridge piers are so arranged as to preserve boat access ramps
that bring tourists into town. The expanded rules may also be meant to
counterweigh environmental benefit against economic hardship.

Certainly, bridges have economic and social as well as environmental
effects. However, a trained social scientist may well doubt that any document
prepared at reasonable cost, in acceptable time, can provide an adequate
answer to social and economic questions. The study may well provide results
inferior to in-context local decisions based on local judgment by elected
officials and local participants about what is best for a community. Nor is it
clear that an EIS is the right place for data informing local debates about the

economy. The expansion of the scope of EIS even risks shifting to economic and social analyses (which are likely to be vastly simplified for the sake of fulfilling legal requirements) the analytical time that should have gone into studying true environmental consequences. The rest of this chapter eschews these accretions to the EIS process. We focus on the core of the process, the assessment of effects that are "environmental" in the ordinary sense.

ASSESSING ALTERNATIVES

When the need to consider every regulation is put aside, and the nonenvironmental requirements are stripped away, the EIS at its core presents a logical and comprehensive procedure for assessing environmental consequences of civic projects. For a given bridge proposal, the EIS attempts, in effect, to catalog all significant impacts.

The EIS begins by evaluating alternative alignments for the bridge. What is more, it considers the possibility that the city would be better off with no added bridge at all. Study of the "no bridge" alternative is especially important in areas where the public's main motive is to reduce traffic congestion. A new bridge would, after all, seem to promise that it will speed up the traffic.

As we have already mentioned, however, much research in transportation planning has shown that, in congested cities, emptier roads and faster travel times make more people decide to travel by car or make them take more frequent or longer trips. This is known as *induced travel demand*. Instead of resolving congestion, the added traffic lanes that a new bridge opens up induce new travel, adding to the sum total of traffic in the area, eventually restoring the congestion the new bridge was meant to solve. For this reason, at the early stages of an EIS, environmentally aware transportation planners should consider *travel demand management* policies instead of a new bridge. For example, faster emergency response (to clear up traffic accidents and clear away disabled cars) can do much to alleviate traffic congestion without the need for construction, and at much less cost.

In Great Lake City the travel demand management option doesn't apply. The city's downtown is not congested. The very purpose of the new bridge to the promontory is to increase waterfront access, land development, economic activity, and recreation around the harbor. Toward that goal (if water taxis are shown to be inadequate for the purpose), there is no good alternative to a new bridge. Early in our hypothetical EIS process, therefore, the agencies conducting the EIS reject the "no bridge" option.

Now EIS planners must consider alternative alignments. It may seem obvious that bridge planners, architects, and engineers would do so, whether the EIS requires them to or not. They do, after all, have to find the location conducive to the most cost-effective structure for handling the projected traffic load. EIS or no EIS, the design professionals have to consider span length, elevation differentials between the river banks, water surface level, bridge types suited to the place, ramp configurations, subsurface soils, connections to existing roads, and land use at the bridge landing sites, among many other factors.

The EIS process makes sure that environmental impacts are assessed as well. Let us say planners have identified for the Harbor Bridge project three possible locations at which structures could be built. From figure 9.1, we can readily observe why Great Lake City officials prefer alternatives 1 and 2 over 3. The former two options are close to downtown and thereby conducive to the primary project goal: increased public access and economic activity around the harbor.

Yet alternative 3 may seem to have advantages. Requiring a shorter span, this alignment allows for a simple, relatively inexpensive, girder bridge. Its two piers are on dry land, saving on construction cost and avoiding ecological disturbance in wetlands. We may reason that lower cost and seemingly lesser environmental effect offset the inferior location. However, let us not be hasty. To have any chance to create the harbor activity that the bridge planners envision, alternative 3 has to be connected to the downtown by the construction of a new road along the south bank of Great River. The new road would increase erosion and would spill more sediment and runoff into the river than either of the other bridge alternatives. And the road heightens flooding risks, unless expensive (and environmentally disruptive) embankments are built. So on balance, even on environmental grounds, alternatives 1 and 2 are superior to alternative 3.

Of these remaining two, alternative 1 spans a narrower part of the waterway, allowing it, too, to be a relatively inexpensive girder bridge. But the pier and landings on the bridge's outer landing (on the promontory) would disrupt a designated wetland. So, should the city prefer alternative 2? At that wider crossing, planners propose a cable-stayed bridge. It must be high above the water, to allow maritime traffic underneath, and the requisite steep rising ramps and added road work also steeply raise the price. More than construction cost, however, environmental impacts cataloged in the EIS process spell this alternative's doom. The EIS reveals that the harbor is a migratory flyway for many bird species, including rare ones. These can smash into towers or stays, a possibility that renders this alternative unacceptable.

Viewing the choices, project managers opt for alternative 1. The EIS process has done its initial job, therefore, just by helping decision makers avoid the alternatives generating the larger environmental impacts. Now, in view of the single selected site, the EIS analysts must continue their work,

cataloging this single project's environmental effects and finding ways of
remedying them.

CATALOGING THE EFFECTS, FINDING REMEDIES

Looking at the maps for the proposed Harbor Bridge (at the alternative 1
site) in Great Lake City, investigators immediately observe that the bridge
pier at the outer landing disturbs a freshwater wetland. Through field investi-
gations to examine soil, vegetation, and hydrology, a wetlands expert should
now report on the extent of the interference and propose ways to remedy, or
in the standard language, *mitigate* it. In EIS practice, the preferred sequence
of activities to mitigate environmental harm is (in simplified form) as follows: *avoid* the effect if at all possible, but otherwise sequentially favor
each of the preceding over the subsequent options: *minimize, rectify, reduce,
compensate,* and *monitor.*

If the bridge structure at the selected site must have a pier in the
wetland (if project managers cannot *avoid* it), it may yet be possible to
minimize effect by reducing the pier's and ramps' footprints or positioning
them in the least sensitive part of the wetland. Even so, disruptions will
be larger during the construction than afterward. Some projects *reduce* the
construction impacts by installing temporary raised platforms and board-
walks on which construction equipment is positioned. The project team
may afterward *rectify* the disturbance by removing those platforms to restore
the wetland habitat and *monitor* the affected area to assure that vegetation
regrows and wildlife returns. If the effect on the wetland is permanent and
unavoidable, the project managers can *compensate*: project investors pay for
the expansion or preservation of a wetland located elsewhere.

The EIS investigators also predict that the bridge, its ramps, new
accessory roadwork, and nearby soils compacted to restrain earth movement near these structures will all release additional rain and snow into
the harbor. This rain and snow is usually known as *stormwater.* Instead
of soaking into the ground (as it does on open soil before the structure is
built), the stormwater washes over these paved surfaces, carrying dirt, debris,
road de-icing salt, motor oil, and other pollutants into the water. Sediment
makes the water cloudier, possibly harming aquatic plants; debris can choke
wildlife and simply litter the water surface; and paint and oil, among other
substances, can poison fish and degrade drinking water.

Sedimentation is likely to be most intense during construction, when
excavation and cutting and filling release silt into the waterway. Also, construction equipment can leak fuels. In keeping with "best management prac-
tices," the project team can make sure that construction contractors *reduce*
these impacts by installing fences or *curtains* in the water to trap silt. The

construction crews should also quickly cover exposed riverbank soils with mulch (to reduce runoff) and keep construction vehicles as far as possible from the riverbank.

For the long run, during and after construction, project managers should ensure that exposed soils are reseeded for vegetative cover, to further *reduce* erosion. But runoff from the added pavement is inevitable. In the winter climates near the Great Lakes, salts used for de-icing roadways or adding traction on snow are the most serious pollutants. It is technically possible to *minimize* the runoff by installing gutters on the bridge and channeling stormwater toward treatment, but such options are expensive.

Some bridge projects have *compensated* by subsidizing improved stormwater management elsewhere on dry land, where greater quantities of stormwater (greater than would spill off the bridge) can be treated at lesser cost. Project managers may also commit to *monitoring* water quality near the construction site, with a promise to add further mitigating measures if water quality declines more than expected.

Some parts of the Great Lakes are migration pathways for birds, some of them rare species. As we have said, Great Lake City's Harbor is indeed a migratory flyway. Let us say that EIS investigators find that the wetland nature reserve on the promontory, near the outer landings of the bridge, serves as an important bird sanctuary, over 200 species having been sighted there, most of them migratory. Rare ones include, say, the American bittern, blue-winged teal, willow flycatcher, and American woodcock. US environmental agencies have documented that both migratory and resident birds crash into tall structures, including towers supporting suspension bridges or cable-stayed bridges. Project managers have *minimized* this harmful environmental impact, possibly avoided it altogether, by eliminating a cable-supported bridge (alternative 2) from consideration. That birds would smash into the sides or railings of a low-elevation girder bridge seems much less likely, but it is perhaps a possibility that cannot be eliminated.

RESIDUAL QUESTIONS ABOUT EIS

Let us quickly add that EIS investigators may well discover more environmental impacts at Great Lake City's harbor. Bridge construction can dislodge deeply buried river contaminants left over from the city's old industrial days. Construction in the river and wetland may also harm fish habitats, though the fish that survive there may be common and fairly uninteresting from an ecological standpoint. We will not consider these and other impacts because we have made our point: that the EIS process forces the project managers to catalog significantly harmful environmental impacts, draw stakeholders' and the public's attention to them, and propose ways of ameliorating them.

What remains to be said is that the process leaves some unanswered questions. As a big technological artifact in the environment, a bridge is, after all, relatively benign. It is a concrete or steel structure that in itself (leaving aside the traffic on it) usually does not pollute. How do we judge whether its extent of environmental impact, say two acres of disrupted wetland, added stormwater spilling from the bridge into the harbor, and a rare woodcock or two killed by a truck, is worth so much bother? Would it be better to disregard these arguably minor impacts, build the project quickly, and put the saved funds toward mitigating more severe environmental harms in other locations?

At the project site, how do we weight the relative harms from wetlands disruption, stormwater discharge, bird deaths, and release of toxic buried sediments? How much worse is it to harm a pair of blue-winged teal than to lose an acre of wetland, or let road dust and grime (washed off with rain and snow) cause water turbidity? And how much should be spent to mitigate one impact compared to the others?

These are questions to which the EIS analyses cannot provide a technical answer. Or at least there is no methodology that can provide diverse stakeholders, having different interests, an answer they would readily consider scientifically reliable. After all, bird impact studies, stormwater studies, aquatic wildlife studies, and wetlands habitat studies are done by different kinds of scientists. Relative harms, and reductions in harm, seem incommensurable—we do not reliably know how to compare them according to one measurable value standard.

Even without providing answers to these tough questions, the EIS does an important job for citizens of Great Lake City. It reveals the environmental issues that require public debate. It provides clarity on the subset of problems for which scientific answers are available. And if properly done, it also exposes the residual subset for which scientific analysis cannot, at reasonable cost or in reasonable time, tell citizens how to decide.

These questions will have to be resolved by precedents from bridge projects in other localities, by negotiations and parrying among stakeholders, by lawsuits initiated by those who think they can get a better outcome in court, and ultimately by political decisions. What is clear is that it is best to have citizens who are informed about infrastructure and knowledgeable about the stakes involved.

WHAT IS A SUSTAINABLE BRIDGE?

As worries about sustainability have swept the design professions, one of the sustained debates has been about the definition of sustainability. We do in the rest of this chapter get to our ideal of a sustainable bridge, but we start now only with this simple observation: sustainable design means more than the reduction of harmful impacts of the kind that the EIS identifies.

For explanation let us consider a sustainable building. When we avoid positioning it in a way that disrupts natural drainage patterns, and take any number of other steps to reduce harmful effects, we cannot yet legitimately claim that it will be a sustainable building. To make that claim, the building should to do more. It may, for example, use nonscarce local materials, orient windows with respect to solar radiation, make use of solar or geothermal energy, reduce dependence on energy from the electrical grid, and facilitate bicycle and pedestrian access.

Architects have embraced a certification method (the so-called *LEED* method, standing for "Leadership in Energy and Environmental Design") that gives credit for such green features. Once a building is certified by such criteria, it can be reliably said to have achieved a high standard of sustainability. As we write, organizations in the United States and Europe are trying to develop similar certification standards for public infrastructure, including bridges.

Since bridges (unlike many buildings) have to undergo EIS review, the contributions that such new certification would make are not immediately apparent. After all, bridges in comparison to other environmental interventions are relatively inert. What is more, the materials commonly used, namely cement and steel, use resources that are among the most abundant and least harmful of industrial materials. Concrete consists mainly of sand, gravel, stone, and water—materials available throughout the world—mixed with cement to glue it together. Concrete made in the United States even contains industrial waste such as fly ash and furnace slag that, if not recycled in the concrete, would be disposed of in landfills.

Steel, too, is made from one of earth's most abundant elements, iron. Upon a structure's disassembly, steel can be recycled into a new structure, and after that second structure's disassembly into a still newer one, and so forth indefinitely into the future, without loss of the properties that make it valuable for bearing loads. Steel, however, is made at fewer locations in the world and must often be transported long distances, often across the oceans, to the place where it is to be used in a structure. Compared to the exotic substances in electronic and mechanical goods, the steel and concrete in bridges are relatively innocuous. If in that respect most bridges already have a good claim on being sustainable, what would make them more so?

Both concrete and steel do have an important property that sparks environmental concern: both contain *embodied energy*. It is an abstract concept: the total nonrenewable energy that was expended for making the material or the object (the bridge) for which the material is ultimately used. The amount of embodied energy is typically measured via the metric system as mega-Joules (MJ) or giga-Joules (GJ) per unit of mass (such as kilogram or metric ton) or unit area (such as square meter of bridge deck).

Much caution is needed in deciding just what exactly is meant by embodied energy for a bridge. Depending on the researcher who measures it, it can include any or all of the first five items on the following list, but almost never the sixth:

1. Embodiment in the material (from material extraction, transportation, and manufacture)

2. Construction of the bridge

3. Physical maintenance and repair of the bridge over its lifetime

4. Demolition and disassembly at the end of standard bridge life

5. Life-cycle embodiment: sum of the above

6. *Minus* Extra life-cycle embodiment (applicable to highly durable bridges only): the energy embodiment avoided because replacement bridges did not have to be built.

According to authors Ashley and Lemay, who summarize several studies comparing embodiment in building materials, concrete contains far less embodied energy than steel, though the differential depends on the specific types used. One study they cite finds that reinforced concrete embodies 2.5 GJ per metric ton, while steel embodied 30 GJ per metric ton. Though the differential is not always that large, it is clear across studies that steel embodies more energy than concrete. Generally, reinforced concrete has more embodied energy than unreinforced concrete, and virgin steel more than recycled steel. For steel, the location at which it is used affects results, since steel is far more likely than concrete to be imported, and for that reason to have more transportation energy embodied in it.

It would seem that concrete is to be preferred, hands down, over steel. But for comparing energy embodied in bridges, the measures just provided are not fair. The structural concrete members needed to support a given load have to have more mass (have to be heavier and fatter) than steel members. So it is best to compare embodiment in the materials per square meter of bridge area (measured by some as deck area and by others as project area). For longer spans, steel will do better because it is more efficient as a structural material. On a cable-stayed bridge, a deck made of concrete will be much heavier than one made of steel, requiring that more materials be used for the concrete towers, foundations, and stays.

The story does not end here. For a fuller assessment of bridge sustainability, the energy embodied in the construction process, in maintenance and repair through the life of the bridge, in eventual demolition or disassembly, and in disposing of the eventual waste should also be measured.

Even more importantly, bridge durability matters. In dry climates, both steel and concrete are very durable. In damper climates, concrete is far more durable than steel, especially if the concrete is properly maintained (as to keep out penetration by chlorine). In principle, a deck-arch bridge made of concrete and well maintained can last a thousand years, as compared to the typical 75 to 100 years for a standard bridge.

CONCLUSION: THE THOUSAND-YEAR BRIDGE

The value of the EIS method is that it catalogs a project's harmful environmental impacts, allows agencies and stakeholders to compare them, and provides options for mitigation of impacts. It provides penalties for a harmful bridge; it does not usually provide incentives for positively contributing to the health of the environment. Broadly defined, a sustainable bridge is one that both minimizes bad effects and contributes to good effects on the environment.

It has good effects in part through the positioning and role of the physical bridge in the community. Say that the proposed bridge at Great Lake City makes attractive residential land available near downtown. If people then settle there, and are there more likely to walk to stores and downtown jobs and make use of newly connected outdoor areas for recreation, instead of settling in suburbs and having to drive to work, the bridge has succeeded in compacting activity and reducing sprawl.

Even more, a bridge has good effects on environmental health if it is a net avoider of the embodiment of nonrenewable resources. Odd as that may seem, that is just what a very durable bridge is. Take as an illustration a concrete deck-arch bridge (the deck rests on top of the arch) that has been shown, from EIS studies, to have little by way of harmful environmental impact. The concrete material in the bridge is not a glutton for embodied energy. Most interestingly, like the Roman arch bridges and aqueducts of old, it can last a thousand years. Over the millennium, it helps avoid the embodied energy from nine replacement bridges with conventional life spans. Though it is oversimplified to say so, there is a legitimate sense in which the bridge makes the environment healthier by saving the energy value of nine bridges. That is the greenest possible bridge.

Further Reading

The Great Lakes City example was drawn in part from the US Department of Transportation Federal Highway Administration, New York State Department of Transportation, and Erie Canal Harbor Development Corporation's *Buffalo Harbor Bridge Final Scoping Report*, Buffalo, NY, March 2010, but several simplifications were inserted, plus geographical features invented to make the case more illustrative. A story of the construction of a dramatic new bridge, with much attention to environmental challenges, may be found in the South Carolina Department of Transportation, *Spanning a River, Reaching a Community: The Story of the Ravenel Bridge*, 2006. Guides to environmental issues and the EIS process for highway infrastructure, including bridges, can be found on the web page of the Federal Highway Administration. That agency also has a good introduction on "induced travel," easily found by searching under those key words. There are several textbooks on EIS, including Michael R. Greenberg, *The Environmental Impact Statement after Two Generations: Managing Environmental Power*, Routledge 2012. Readers who want to learn more will find it worthwhile to inspect an EIS for a project in their own locality.

Information on the LEED architectural standards may be found on the web page of the US Green Building Council. As we write, the Institute for Sustainable Infrastructure (formed by the American Society of Civil Engineers, American Council of Engineering Companies, and the American Public Works Association) is preparing certification standards for infrastructure. On concrete's "eco-friendly" features, see Erin Ashley and Lionel Lemay: "Concrete's Contribution to Sustainable Development," *Journal of Green Building*, Vol. 3, No. 4, 2008, pp. 37–49; this article has references to many additional studies. Also see M. Myint Lwin, "Sustainability Considerations in Bridge Design, Construction, and Maintenance," *ASPIRE: The Concrete* Magazine, Winter 2007. Possibly in defense against pro-concrete trends, the British Constructional Steelwork Association has issued a report titled "Steel: The Sustainable Bridge Solution" (undated, apparently 2010).

Bridge engineers who are investigating the value of concrete in bridge sustainability include David Long, "An Environmental Comparison of Bridge Forms," *Bridge Engineering* [Proceedings of the Institution of Civil Engineers], Vol. 159, Issue BE4, 2006; A. E. Long and colleagues in "Sustainable Bridge Construction Through Innovative Advances," *Bridge Engineering* [Proceedings of the Institution of Civil Engineers], Vol. 161, Issue BE4, 2008; and an unpublished paper by Andy Wong, "Sustainability of Infrastructure Projects in Hong Kong."

TEN

DELIVERING THE BRIDGE

Contrary to what many expect, the most complex part of making a bridge is not the structural engineering, but the process of *project delivery,* by which the bridge progresses from an idea to a constructed object. This process occurs against a background in which public support has to be assured, disagreements (such as lawsuits) resolved, laws obeyed, environments protected, contracts let, materials and labor assembled, and moneys appropriated. For the various decision makers and professionals, the overarching challenge is that of efficiently bringing together these multifarious tasks to finally build the bridge.

Will the structure's design be as hoped, or will it have to be trimmed to meet budget? Will the architectural effect be as dramatic as the public expects? Will the structure and attached public works not only avoid harming the environment, but actually enhance it? Will the project be done on time? Will it come into existence at anticipated cost, or will it have cost overruns? In view of the expense and complexity of multiple layers of review, will it be built at all, after a decade of study? The answers will emerge over many years through the sequenced webs of activity through which the project moves from concept to delivery.

BRIDGE PROJECTS SINCE THE FREEWAY REVOLTS

Ever since the birth of modern civil engineering, construction of a major bridge has been a complex and painstaking dance, taking years to complete. Unexpected events, such as a labor dispute, or a storm that damages an incomplete structure, can escalate costs and delay completion. But, in our time, construction is rarely the critical component that encumbers the project. Rather, increasingly since the 1960s, major public works projects have been subjected to ever more layers of environmental studies, community or interest-group review, and legal challenge. Bridge engineering tasks have

137

137

been submerged under ever more elaborate regulatory procedures—often for good reason. In short, the making of bridges has evolved from a process dominated by engineering and architectural considerations, to one managed under fantastically complex regulations, court rulings, budget cycles, and relationships between government agencies.

This evolution has occurred in response to real havoc that infrastructure projects have caused, though the culprit has more often been the highway than the bridge. Through the 1950s and 1960s, newly built highways sliced through numerous urban neighborhoods where they were not appreciated, and by the mid-1960s, in some places, positively hated. Disliking the noise, the demolitions, and the cuts into the fabric of the city, community groups organized to fight new highway extensions—a phenomenon now known as the "freeway revolts." Whether because of racial prejudice or the lower cost of land-acquisition in poor neighborhoods or both, many a highway ran through poor African-American sections of cities, tearing neighborhoods apart, or placing a barrier between poor minority neighborhoods and wealthier ones. Some highways also damaged pleasant natural environments or destroyed historic buildings.

By the late 1960s, the groups that arose to fight highways often succeeded. Throughout the United States, there still are stretches of highway that lead to dead ends, or multilane limited-access highway segments pointlessly placed between minor roads. These are the remnants of community revolts and budget reconsiderations that terminated highway projects in midstream. There were for a while even a few ramps and bridges (and partially built bridges) to nowhere. In the same decades, Americans also became aware of the environmental effects of highway infrastructure: the cuts through forests, the alterations of stream beds, destruction of animal and plant habitats, and induced erosion, among other impacts.

The new regulations that have emerged since 1969 reflect both political trends: urban community activism and environmentalism. In ever more intricate accretions, the new rules have mandated environmental impact reviews plus opportunities for public comment by community groups, interested parties, and political activists. Though bridges are far less likely than highways to cause bad effects, bridges are of course inseparable from highway systems, so they have come to be regulated under the same laws.

In America, the key law is the National Environmental Policy Act of 1969, commonly known as NEPA. It mandates that, for projects that use any federal funds, officials must consider environmental factors, and even economic and social factors, not just budgetary and technical ones. Most notably, the law created the *NEPA Process*, the sequence by which major projects have to undergo a series of studies and hearings leading to a final environmental impact statement (EIS, described in the previous chapter).

Through subsequent amendment and refinement of this federal law, and the adoption of copy-cat laws in various states, the NEPA Process has come to dominate *major* public works. (Minor projects and projects with no expected deleterious environmental consequences are exempted.) In the development of a bridge, the sequencing of project stages is so thoroughly shaped by the NEPA that it is often impossible to distinguish steps meant specifically for bridge engineering and design from steps meant for dealing with environmental (and cultural and economic) impact.

The environmental regulations and the requirements for citizen involvement have done enormous good for the United States. They have prevented many poorly conceived and environmentally harmful public works. But when is a good idea taken too far, so far that it results in delays, cost overruns, stagnation, and harm to communities? This is a question to which we return at the end of the chapter.

FROM PROPOSAL TO COMPLETED CONSTRUCTION

As a result of NEPA and its progeny, bridge development has become a process managed by multidisciplinary teams, responding to complex regulations, usually under the authority of a state department of transportation, but interacting with many other agencies. The full-fledged process to be described here applies to construction of new bridges or replacement of old bridges—projects that have significant environmental impacts. The process is typically understood in stages. Though NEPA terms are shared, state transportation agencies do not necessarily call the stages by the same names, or divide them at the same rungs. Terminology differs somewhat from state to state.

The description we offer here is a simplified amalgam of the processes used in New York State and in Oregon. In our version, the bridge development project has six stages: (1) initiation, (2) scoping, (3) preliminary design and environmental review, (4) detailed design and agreements, (5) bidding and contracting, and (6) construction. Put this way, the process is linear and ends at the completion of the bridge. Thought of differently, the process is a recurring cycle: after construction comes operation and maintenance, which is followed by rehabilitation, which is eventually succeeded by decommissioning or otherwise by replacement or reconstruction, when the process starts all over again. However, in this chapter we discuss just the six stages that conclude when construction is complete.

We use examples from the development of two bridges. The first is a pair of new deck-arch bridges over Cattaraugus Creek as part of the Route 219 Expressway in upstate New York, completed in 2011. The second is the replacement of the Kosciuszko Bridge, part of the Brooklyn-Queens

Expressway in New York City, its future type not yet decided when we wrote. (See figure 10.1.) The first is part of a new stretch of limited-access highway on a new alignment in a rural area; the second is a reconstruction on an old overburdened highway amid an industrial landscape in the nation's largest city. As we shall see, they offer different lessons about the bridge development process.

THE SIX STAGES

Stage 1: Project initiation

Unless it is a simple reconstruction or replacement on an existing highway, the proposal to build a bridge is likely to have many origins, hard to trace. It may have been promoted at various times by citizen groups, local elected officials, state legislators, and local governments. Towns, cities, and counties may push for a bridge. They may have identified it in plans for capital investment: as compared to water and sewer repairs, street work, or a new city hall, the bridge may have turned out to be a town's higher priority.

By contrast to a local municipality, which must prioritize all kinds of public works, a state department of transportation (DoT) examines only transportation projects. The agency's project choice may emerge from the database by which it keeps track of the conditions of its bridges. With it, the agency can identify the ones most in need of repair or replacement. The DoT may also pick projects based on a long-term transportation plan. Over a 20-year period, the plan should project traffic demand and road capacities, including bridges and ramps, helping in the choice of places where new bridge capacity is needed. (These plans are also made by a separate kind of agency, namely the *metropolitan planning organization*, which we describe below.) If asked, agency officials are likely to point to such technical systems as the origins of their project proposals. For large projects, it would also be normal for elected officials to try to influence the decisions. Such influence is exerted at the highest levels, on agencies usually led by political appointees, so it is difficult to disentangle the technical from political influences on project choice.

For an envisioned north-south highway that would pass over Cattaraugus Creek in upstate New York, discussions and studies can be traced as far back as the 1940s, when it was already realized that the city of Buffalo, which is on the Canadian border, lacked a direct high-capacity highway to the south, toward Pennsylvania and Washington, D.C. It was thought that such a link would improve Buffalo's economy and facilitate trade with Canada. For half a century, local leaders and representatives in the US

Figure 10.1. *Upper:* The twin-span Cattaraugus Creek Bridge, carrying Route 219, an expressway in upstate New York. It is pictured in October 2010, just prior to opening, with the old Route 219 bridge in the background. Photo by Frank O'Connor. *Lower:* A view of the Kosciuszko Bridge on the Brooklyn-Queens Expressway in New York City, taken October 2010. Photo by Steven W. Bennett, PE.

Congress advocated for the new expressway's funding. Numerous studies examined traffic demand and potential alignments.

As in this case, highway proposals may be discussed, studied, debated, rejected, and studied again for decades. Such efforts eventually result in an *initial project proposal*, formally produced within the state DoT. The proposal gives the justifications for the bridge, the following being the most common: to preserve infrastructure, relieve congestion, increase capacity to meet growing traffic, and promote economic development. The proposal also anticipates environmental challenges, suggests a preliminary project schedule, and roughly estimates costs. It also anticipates whether the project will formally be termed "major." If it is, it sets in motion the most elaborate NEPA process requirements.

For major projects, this proposal is barely the first step. It does not intend to actually get shovels into the ground. Rather, it asks for permission for the next stage in the process, the more expensive "scoping" study, for which expense runs into the millions per project. For a DoT to accept the proposal means that the agency is ready to invest seriously in study. Acceptance marks the end of stage one. In calculating the time it takes to develop a bridge, it is best to begin at the end of stage one, with the decision to do scoping studies. That is when the clock starts.

Stage 2: Project Scoping

Since a federal highway law enacted in 1962, multicounty American urban regions have had to collaborate for transportation planning through entities generically known as *metropolitan planning organizations*, though each has a different local name. Obviously, transportation continues across municipal borders, so these organizations have the role of coordinating municipalities' road and transit (and more recently, pedestrian and biking) plans.

Almost all these organizations are overseen by delegates of elected officials and agency leaders from the region. In typically decentralized American government, in which higher-level elected officials (say governors) cannot not give direct orders to lower-level elected officials (mayors or county executives), the metropolitan planning organization (MPO) exerts its influence through the power of the purse: federal moneys do not get spent unless the organization approves. Sooner or later, proposals to spend federal transportation money in the region must pass through this organization.

The various local governments and local agencies that jointly direct the MPO may make proposals to it for small projects, such as road repairs and bike paths, and large projects, such as rail projects, road widening, and bridges. They hope to get their proposal on the list of projects approved for federal funding. For a major project, there must be a full scoping study before it can get on the list. The agency responsible for the scoping is likely to be the state DoT. At the time we write, costs for scoping a major bridge project easily exceed $2 million per year.

To begin scoping, the DoT appoints a multidisciplinary team to oversee the project. Specific objectives have to be stated (safety, mobility, economic development) and supported with documentation (accident rates, highway usage, economic conditions). For a new highway project such as that going through Cattaraugus County, potential alignments have to be identified. In this county, there were 27 plausible alignments compared to the "null alternative" of doing nothing and the additional alternative of just upgrading existing roads. Since the routes going through the county are part of a larger north-south road network affecting many states and even cross-border travel (to Canada), researchers had to conduct a *corridor study* of travel demands and road supply in the broad area that the new highway would serve.

In upstate New York, the purpose was mainly economic development. For the 70-year-old Kosciuszko Bridge in New York City, the objective was to respond to deterioration, which had been documented through repeated inspections. An additional purpose was to reduce traffic accidents and increase mobility on the Brooklyn-Queens Expressway, if *mobility* is the right word, since the expressway often resembles a lengthy parking lot.

The scoping study must identify potential project impacts, both positive and negative. As the scope of environmental review has expanded since 1969, the impacts this study now must consider include those not just on lands, waters, flora, fauna, ecosystems, and pollution, but also on archeological sites, historic buildings, and community character, among other matters. The study must also estimate costs and give a preliminary construction schedule. The process must be publicly announced, and members of the public, especially stakeholders having an interest in the project, must be given the opportunity to question, comment, advise, and oppose. The outcome is a thick scoping report, made available to the public and sent to relevant agencies and governments, and to the MPO.

The MPO maintains a multiyear projection of projects it supports. It is a fiscally constrained list: it anticipates the amounts available to be spent and projected future amounts. Called the *transportation improvement program* (TIP), the list should have no more projects than will reasonably be funded in the next few years. When the scoping project is complete, the MPO decides whether the candidate project gets on the TIP. More projects are scoped than get on the TIP—there is never enough money.

The TIP rosters from all the regional lists must be pulled together to make up the statewide transportation improvement program, the *state* TIP. (For those who want to keep up, that's called a *STIP.*) There is no guarantee that a project on the regional TIP will get on the state TIP. Getting on the state TIP is the achievement that ends the second stage in the process. By now, leading transportation officials, including the commissioner of the state DoT, are on record in support of the project. The project is henceforth scheduled to receive funds for construction—though it actually gets built only if all the subsequent steps in the process go well. A project on the state TIP has a high chance of being constructed, though large, complex, and controversial projects naturally face more challenges to completion than minor ones.

Stage 3: Preliminary Design and Environmental Review

For the next stage, the DoT again appoints a multidisciplinary project team and project leader, perhaps some of the same professionals who previously worked on project scoping. The team then assigns staff and consultants to a series of studies, whenever possible recycling data already assembled during scoping.

For Route 219 through Cattaraugus County, the project team had to study impacts of the most promising alignments. Here is a sampling of the issues studied: impacts on road access to villages along the route, community character, planned housing, oil and gas wells, agricultural soils, agricultural districts, hazardous waste sites, aquifers, archeological sites, state forests, churches and schools, and historic buildings. Effects had to be examined on hundreds of streams, most of them rivulets. Their combined roles in watersheds and effects on water quality also had to be understood. Since the expressway would pass over Cattaraugus Creek, a scenic gorge, there had to be intensive study of migratory birds in the gorge and of habitats for the clubshell mussel, a species listed by the US Fish and Wildlife Service as endangered. From each candidate alignment, there were separate implications for acquisition of land rights-of-way, whose numbers and costs of acquisition had to be estimated.

In general, such a team produces two interrelated reports. One is the *draft design report*, which proposes the engineering design for road, embankments, culverts, bridges, etc.: the types of information that will be needed for construction. The other is the *draft environmental impact statement* (EIS). The documents are then distributed to other agencies that will have a say in final permitting. The draft EIS is also made available to the public, so individual citizens and groups can make their views known in public hearings and meetings. Based on comments received and problems identified through public involvement, and after any additional studies are conducted to address these problems, the team issues a *final design report* and final EIS. If there are further objections and lawsuits, the reports may turn out to be not so final after all. Further studies are then conducted, further comment sought, and further hearings held, to prepare a further report, which may indeed be final. It is not uncommon for some team members to have retired by then, to be replaced by new ones.

In response to each local project, citizens groups and specialized stakeholder groups arise to make their feelings known. In Cattaraugus County, their concerns included protection of historic farmhouses, endangerment of ecosystems, highway access for businesses, and the inevitable geographic splits in communities that limited-access highways cause. In New York City, topics of concern included traffic disruption during construction, construction noise, air pollution, dislocation of businesses, and remediation of the polluted creek that the bridge spans.

In the New York City's Kosciuszko Bridge project, the state DoT project team established two advisory committees. One consisted of local residents, representatives of community groups, and elected officials. The other was made up of technical representatives from agencies that would have to give permits for the project. These included the U.S. Coast Guard, the Army Corps of Engineers, the state Department of Environmental Conservation, and various New York City agencies dealing with streets, parks, and the environment.

There were 150 meetings, some 35 for the community advisory group, 20 for the interagency group, and dozens of miscellaneous other open houses and hearings. In response to public objections, the state DoT promised to put a bikeway alongside the Brooklyn-Queens Expressway, remediate a hazardous waste site that happened to be nearby, and build boat launches on the creek. Similarly, at the Route 219 Expressway in upstate New York, the DoT responded to concerns by committing to improved aquatic and wetlands habitats near the selected alignment, restoration of a degraded stream, establishment of wildlife viewing areas, and a shift of the road alignment to save historic farmhouses.

In all, the basic design emerged from a long process of give and take. As should be clear, design of the bridge and road has come to be barely distinguishable from design meant for environmental adjustment. Structural design, alignment, and environmental impacts had to be iteratively considered until the design report and EIS were complete and consistent with each other.

To understand how stage three ends, let us recall that the federal legislation called NEPA profoundly shapes this entire process, because it requires agencies spending federal funds to certify that environmental impacts are acceptable. For highway and bridge projects, federal funds run through the Federal Highway Administration. The state DoT submits to the federal agency the final EIS. If it is acceptable, the Federal Highway Administration issues a positive *record of decision*. This is the end of stage three.

From the beginning of scoping in about 1992 (procedures were different then, so the start date is not that clear) through issuance of the record
of decision in 2003, the second and third stages for the Cattaraugus County
Route 219 Expressway segment lasted about eleven years. For Kosciuszko
Bridge, the second stage lasted from 1999 to 2002, and the third from
then through 2009, the year in which the record of decision was issued,
for a combined total of over ten years. The issuance of that decision tells
engineers that they can now stop considering further alternatives and focus
rather on detailed design.

Stage 4: Detailed Design and Agreements

Once an alignment has been chosen and environmental concerns met,
detailed design refers to final engineering to prepare specifications sufficient
for construction and for regulatory permits. In the New York State DoT,
detailed design is expected to create documents that make the project 90
percent ready for execution.

At the start of this stage in the process for the Route 219 Expressway,
the alignment that had been identified was still just a 500-foot-wide swath.
During detailed design, the precise alignment within that corridor had to
be specified. Once the right-of-way was exactly defined, the DoT could go
ahead and acquire the designated properties.

Upon careful mapping and analysis of the alignment, plans had to
be prepared for pavement, retaining walls, the soil subsurface on which
the road rests, erosion control, signs and signals, road illumination, water
management, sourcing and disposal of construction materials, and culverts
and bridges, among many other matters. The plans had to be reviewed for
practical constructability. Not least, the project team had to develop cost
estimates and a construction schedule.

The detailed design concluded with the completion of a document
known as *plans, specifications, and estimates* (PS&E). It must be specific
enough to be bid upon and to facilitate construction.

Overall, this fourth stage takes one to two years. The critical difficulties arise less from the practice of engineering design than from other
activities that must take place concurrently. A specialized office of the state
DoT obtains land owners' agreements to have their properties acquired—far
more of a challenge for a new highway alignment in upstate New York than
for a replacement bridge in New York City. Staff members contact the land-
owners and offer prices based on standard calculation. In hardship cases, as
when a farmer's property is split in a way that ruins the business, the agency
can offer to purchase more of the land than is essential for the project. In
New York State, if the owner refuses to sell, the DoT sends the case to the

office of the Attorney General, where proceedings may begin to acquire the property through the legal procedure known as *eminent domain*, in which the owner can be forced to give up the land at a price determined in court.

Concurrently, DoT staff has to gain agreements from utilities to relocate water and sewer pipes, power lines, gas lines, and other infrastructure with which the highway would interfere. As may be imagined, the utilities do not always respond with alacrity. Further, the DoT must obtain permits from the various state and federal agencies overseeing endangered species, water quality, navigation, air quality, watersheds, and other matters. Though a few land acquisition disputes and utility relocations may linger, stage four ends when the PS&E is complete and when permits are signed.

Stage 5: Bidding and Contracting

This is the shortest of the stages in project development, taking up just a few months, but is nonetheless essential. As this stage starts, the project team that has run the process so far shifts responsibility to the office that manages the letting of projects for bid. The new office reviews the PS&E to make sure the document meets standards. Even at this stage, the DoT can stop the letting process if state funds are deemed insufficient.

The letting documents are electronically posted and made available for bidding, in a process with which those experienced in the highway trades are well aware. For major projects, the bidders are general contractors who plan to build the project, usually by a combination of in-house work and specialized subcontractors. The bidders must follow precise guidelines, including the posting of a bond. After bids are in, staff reviews them for compliance and makes a recommendation. Occasionally, delays occur when a losing bidder protests or sues, or when the winner is found to be noncompliant with a regulation, such as a requirement to hire minority contractors. Barring such problems, the award is made and a contract executed. The signed contract marks the end of stage five.

Stage 6: Construction

The general contractor supplies materials and performs the construction directly or through subcontractors. The DoT in turn has to make sure that the work complies with the contract and specifications. The agency does so by assigning staff to monitor and document progress, including quality and workmanship. To assure that the materials used meet contract specifications, the agency sends samples for analysis at a certified laboratory. DoT environmental inspectors have the additional job of overseeing environmental compliance; other environmental agencies visit to perform further inspections.

DoT inspectors must investigate discrepancies between work performed and work initially specified. When a dispute arises, the DoT's project manager (sometimes known as the engineer-in-charge or the resident engineer) must first attempt to resolve it amicably, but may in rare cases, if the matter continues to be contested, send it to third-party dispute resolution. A challenging matter in construction is the *change order*. Such orders are made when on-the-ground conditions prevent construction in the manner envisioned in the original design. DoTs have elaborate procedures by which to evaluate and approve (or disapprove) change orders. The more expensive the change request, the higher it must be sent within the agency's hierarchy to receive approval.

We cannot yet describe construction to renovate the Kosciuszko Bridge, because the project is still in the detailed-design stage; construction will not begin for years. What we do know is that one aspect will be far more complicated than in upstate New York. During bridge construction, six lanes of traffic will continue to creep on the Brooklyn-Queens expressway, on temporary ramps and bridge structures constructed for the purpose.

For the Route 219 Expressway in Cattaraugus County, funds were received to build a 4.2-mile segment. The contract was awarded in January 20007, with completion initially set for July 2009, at an estimated cost of $86 million (rounded off). But serious problems appeared, first in the form of cracks.

As project engineers have since explained, the problems arose from five layers of subsurface soil under more than a mile-long stretch of the new road. The construction had unanticipated effects on these soils: sliding processes ensued under the surface, causing cracks in the embankments. Experts who were brought in discovered that the soils had been put under intense water pressure; the pressure had to be reduced to make the road viable. Change orders permitted the contractor to do the extra work of installing drains at frequent intervals through the soil to allow water to emerge under the road. And embankments had to be reconfigured. State DoT approved changes of over $40 million, for a final project tab of $126 million.

The project was completed in July 2011. Traffic is now flowing over the highway, and beautiful twin-arch bridges grace Cattaraugus Creek. If the starting point is taken as the beginning of scoping studies, roughly in 1992, then the whole project took nineteen years.

THE PROJECT DELIVERY PROBLEM

Let us now recap the process by which a major bridge project is built, remembering the important word *major*. The word refers in part to cost but also to environmental effects. A small bridge with significant effects on a stream should undergo full environmental reviews and be labeled "major," while a larger and more expensive overpass on an established highway alignment in a paved-over setting would perhaps (there are other considerations) be exempted from environmental review.

Viewing the delivery of a major project, most observers will have similar reactions: it takes a very long time. A remarkably large portion of the effort (time and expense) goes into scoping, preliminary design, and environmental review stages. Projects are prone to cost overruns. Yet this painstaking attention to process has brought the United States many advantages.

Table 10.1. Stages in a Major Public Projects in the United States

Infrastructure Type (new build, expansion, or replacement)			Bridge	Highway
Stage	Conclusion of the Stage	Years	% of Project Cost	% of Project Cost
1. Initiation	Acceptance of proposal (to begin scoping)	n.a.	10.6*	18.2*
2. Scoping	Listing of project on state Transportation Improvement Program	1–2		
3. Basic Design & Environ. Review	Final Design Report and Final EIS; Record of Decision	2–10		
4. Detailed Design & Agreements	Plans, Specifications & Estimates (PS&E), rights-of-way, utility relocations, permits	2–3	6.7** 1.3***	6.7** 3.4***
5. Bidding & Contracting	Execution of contract	0.5	0.1	0.1
6. Construction	Completion and approval of project	2–5	81.3	71.6
Totals		7.5–20.5	100.0	100.0

Sources: Oregon Department of Transportation and specific NYS DoT projects. Cost data courtesy of Oregon DoT, provided courtesy of Bruce V. Johnson, P.E.
*Overall planning & design, not including detailed engineering design
**Construction engineering (detailed design)
***RoW and utility relocation

Without such careful oversight, the country would surely have many expensive on-the-ground projects that are causes of regret. The long time-line and extensive review in the delivery process give opportunities for second thoughts—for altering, scaling back, or abandoning the project. There are fewer white elephants than there would have been without such review. The careful review has also, we assume, reduced the proportions of the natural landscape covered with artificial surfaces. Fine adjustments in alignment and structure have avoided harm to sensitive lands and ecosystems, preserved rare habitats, saved historic structures, and helped make cities more livable.

This environmental success has come, however, at a cost, and not just the direct costs of the projects themselves. Lengthy review does forestall many harmful indirect consequences, but can also cause others. Consider the lists of projects on transportation improvement programs. These are of given length, because they are constrained by the available budget. As infrastructure projects become more expensive, but overall public works budgets do not grow, the lists necessarily become shorter: from the tail end of each list, projects have to be dropped.

Nor is it always the case that the no-build option helps the environment. At Kosciuszko Bridge, slow traffic spews pollutant into New York City's air, possibly more so than fast traffic, so the wait for the new bridge may in itself be causing environmental harm, while wasting millions of travel-hours. What is more, bridges tend to be bottlenecks because (other things being equal) road construction alongside a gap is less costly than construction across it. Tollbooths also cause bottlenecks. Depending on the site, a new bridge can improve the environment by reducing bottlenecks and hence reducing durations of car operation. When well positioned in a dense urban area, and well endowed with pedestrian and bike lanes, the bridge can also encourage non-motorized travel.

We must not forget that questions about project delivery gain special urgency when the United States' economy grows more slowly than its competitors'. China in particular comes to mind. Sometimes with disastrous environmental sacrifices, China is building roads, railways, great expositions, and new cities at a rapid clip. The Chinese people and their natural environments will experience the harms that this haste brings. Americans should not give up on the transparency and public involvement that the NEPA process launched. Nor should Americans compromise on careful legal processes in land acquisition, nor impose top-down centralized control over municipalities and utilities.

Yet, without sacrificing the values that this careful project delivery process has brought, we should also seek ways to improve it. At a time when national infrastructure is deteriorating and America must continue to compete in the world, it is important to ask whether project delivery can be made more efficient, without net loss to urban environments, natural environments, and the economy.

PROSPECTS FOR IMPROVING PROJECT DELIVERY

In the mid-twentieth century, a single public official, namely Robert Moses, was responsible for building more of New York State's public works, including bridges, than anyone else in the state's history. According to his biographer, he would often underestimate a project's costs to get the state legislature to approve it, but once construction was underway and running out of money, would put the legislators over a barrel—force them to appropriate more funds to escape blame for an unfinished project. To this day, around the world, major projects, especially vast "megaprojects," typically run late and exceed initially estimated costs.

This phenomenon has prompted the speculative hypothesis that those who initially push for the project (and the private parties that benefit from building the project) purposefully underestimate costs. Allegedly, they do so out of personal enthusiasm for projects or because of political or economic pressure, though the channels of this alleged influence on cost estimators remains elusive. While we do not have the evidence by which to fully judge this or other explanations for project overruns, we can say, with a view to the modern-day US delivery process we have described, that this hypothesis is not plausible.

The many varied professionals who estimate project costs, at stages from project initiation through the PS&E, and for various subsets of a project, do not have the incentive to be wrong in their estimates. They have no obvious motive for dishonesty—the estimators gain no eventual benefit from cost overruns or delays. Nor do the public officials who oversee decisions at each stage have a clear motive to underestimate costs. In some respects they are under the contrary motive. Transportation improvement programs impose fiscal constraint. Projects that go over budget eventually subtract from the sum total of projects that can be funded. In modern US project development, officials who want regional projects to move through the approval procedure should, in terms of incentives they face, want projects correctly estimated, so that other favored projects are not later subtracted from the bottom of the list.

A second hypothesis refers to the standard procurement method, in which design is separated from construction: state entities design the bridge and then, after a bidding process, a private entity constructs it. The hypothesis is that this separation creates dissonance between design and construction, causing overruns and delays. The solution often proposed is the *design-build* procedure (which has several variations). In this scenario, a private firm (or joint venture) is contracted for both design and construction. Potentially, there are great advantages. The company knows its particular capabilities in construction and can better align design with construction practice. Since the company is responsible for both designing and building the bridge, it has a powerful incentive to design carefully enough in early stages to avoid construction-stage change orders that it would have to charge to itself. In some cases, construction on some parts of the project can begin even before detailed design is complete, shortening the duration of the project.

Design-build as an infrastructure delivery method was common in the United States a century ago, especially during the canal era. It is again in use in many states, with varied claims on efficacy. Each large project is complex and unique; it may be that conventional bid-based delivery is more adapted to some projects, and design-build to others, but we do not know which method works better under which conditions. The evidence must still be collected. What is clear to us, though, is that design-build cannot cut short the complex process we have described. The private design-build firm and accompanying public authorities must still carry out the environmental reviews, public hearings, repeat studies, right-of-way acquisitions, utility relocations, and permits that slow the project's pace.

Still another hypothesis is that costs escalate and delays occur because of technical errors in estimation. Put baldly in this way, the hypothesis seems unreasonable: it would seem that errors could run in either direction. In response to erroneous estimates, major projects should be as likely to be underbudget and fast-paced as overbudget and slow.

A better way to state this hypothesis is that estimates are subject to uncertainty. There are uncertainties in interagency relations (how fast a utility responds to a relocation request), environmental processes (subsurface soil conditions encountered during construction), and community views (how adamantly a group seeks to save farmhouses from the wrecking ball), among many other project features. The more complex the project, the greater the number of uncertainties faced during estimation. And the more the project is assessed based on its systems of consequences (its linkages to environmental effects), the greater the number of cascading downstream system effects that an unexpected problem (such as soil slides) can set off. But this explanation, too, is speculation on our part, and should be subjected to research.

The fourth and final hypothesis is the obvious one: that complex rules and procedures have slowed down projects. Ever more elaborate expectations placed on projects increase the likelihood that disruptive problems will be found. That leads us to the most difficult question: are all the rules and procedures necessary for achieving the environmentally protective purpose?

If a need is obvious from the start (the need to replace the old Kos-ciuszko Bridge), why should it be necessary to study the need in depth? If it is inevitable that the bridge will have to be replaced, why hold up scoping and detailed design for environmental agreements (to build a boat launch) that can be left for a later stage in the process? Should all activist groups that find fault with a project be given equal power to slow down a project that has broad public benefit, or do some groups have more legitimacy than others?

These are difficult questions, ones we cannot come close to answering here. What we do know is that, in the United States, for bridge projects as for other kinds of public works, project delivery is the greatest single challenge. To continue to meet the nation's infrastructure needs, policy makers, planners and engineers will have to learn how to deliver projects more cost-effectively and faster.

Sources and Further Reading

On the rise of opposition to highway projects, see Raymond A. Mohl, "Stop the Road: Freeway Revolts in American Cities," *Journal of Urban History*, 30: 2004, 674–706. For more on NEPA and subsequent legislation, check the website of the President's Council on Environmental Quality at http://ceq.hss.doe.gov/welcome.html. Information on New York State's stages for managing a transportation development project is in the "Project Development Manual," found on the state DoT's website. On the Kosciuszko Bridge we also consulted environmental impact documents and benefited from a fine 2012 Columbia University master's thesis in urban planning, "Why Transportation Mega-Projects (Often) Fail," by Victor S. Teglasi. On project delivery methods, the basic source is John B. Miller, *Principles of Public and Private Infrastructure Delivery* (Kluwer Academic, 2000); also see more recent articles on the subject by Michael J. Garvin, who also cites extensive additional literature. Regarding the problem of cost-overruns, Bent Flyvbjerg and colleagues report on a multinational study of megaprojects in "Underestimating Costs in Public Works Projects: Error or Lie?" *Journal of the American Planning Association*, 68:3, 2002, 279–295, in which they speculate that estimators purposefully mislead—we take a dissenting view in this chapter. Other recent findings on the matter are in Matti Siemiatycki, "Academics and Auditors: Comparing Perspectives on Transportation Project Cost Overruns," *Journal of Planning Education and Research* 29:1, 2009, 142–156; and in Joseph Sturm and colleagues' "Analysis of Cost Estimation Disclosure in Environmental Impact Statement for Surface Transportation Projects," *Transportation* 38, 2011, 525–544.

PART IV

CONCLUSION

ELEVEN

A BRIDGE SPANNING A MILLENNIUM

We live in times when children and many adults have learned to think that the world's fascination resides in shiny screens. Life's excitement, they think, is in ever newer handheld gadgets or in the newest functions available in hyperspace. The bridges and other large objects that make up our public infrastructures reside, however, not in cyberspace but in ordinary space, the space that human beings have always experienced. To the eye accustomed to the little screen and uneducated in the meanings and value of the built environment, these great artifacts seem staid, seem to just sit there.

Yet, as we have learned in this book, bridges are far from being stagnant entities. They span a gap because they are carefully designed to balance forces of resistance against the dead load constituted by the structure itself, and against the loads imposed on it by traffic and natural events. The forces are exerted through basic processes of compression, tension, shear, bending, and torsion, but in infinite quantitative combinations. Whether a bridge is supported with girders, trusses, arches, or cable stays, it stands as a carefully designed structure, dynamically balancing imparted load against structural resistance, in keeping with constraints of materials and of site. When it is well enough made, it qualifies as structural art. Well made, the bridge can serve both as practical asset and as monument.

Good decisions about a bridge are made when more citizens are aware of the constraints and creativity that go into its making. As we have seen, some basics of bridge engineering are accessible to those who never would have considered the study of the subject. In the eyes of those who have achieved even a basic appreciation, the infrastructure becomes more interesting. For a few, bridge appreciation can even be an avocation, sort of like bird watching. For some readers of this book, the planning and engineering of bridges (and other infrastructures) can even become a fulfilling career. If that happens, we have done part of what we have hoped. But most of all, we have hoped that our book will also assist citizens and public leaders in taking part in bridge decisions for their communities.

As we have seen, a bridge comes to the fore in public discourse when one is thought to be needed in a place that hasn't had one before, or more likely, where one is deteriorating or obsolescent. We estimate that 20,000 to 30,000 such bridge discussions go on each year in the country. Public appreciation of bridges is needed because there will be even more bridge debates to come.

The reason is that a spurt in bridge construction occurred in America in the 1960s and 1970s. Many bridges survive in good shape because of careful, often innovative maintenance over the years. But other bridges are becoming old enough to be infirm and troublesome. Many long-span bridges from that period are accumulating problems. Some can be rehabilitated at great expense; others will have to be torn down and replaced. The bill is coming due for the short lifetimes for which the bridges were built. That we have to face this expense now, just 50 years or so after the ribbons were cut, reveals flaws in the very ways in which we make decisions about expensive infrastructure.

We are shortsighted in part because of cost-benefit methodologies that discount the value of a bridge according to the time-value of money at the time the bridge is being planned. At first glance, such methodologies seem sensible enough. If the bridge is just a financial investment, like stocks, it is to be evaluated in terms of potential business income.

But is a bridge just an investment like other investments? We ask this seemingly naïve question with full understanding that money to be spent is limited and there are no magic sources. Infrastructure is expensive, and choices must be made with care. But it is not at all clear that the discounted excess of benefits over costs is the best way to make them. Benefits and harms for traffic are somewhat amenable to analysis. But benefits and harms for the economy, for heritage, for recreation, for environment, and for identity? Different analysts get different results. One reason to be wary of discounted cost-benefit analysis in the choice of bridges is that the methodology leads to the construction of short-lived bridges, when it is not at all clear that the method properly accounts for the full public interest in the bridge.

Let us make an unremarkable statement: bridges are in places. Often they are in cities, where people conduct commerce and move about. For them, the bridge is a practical asset. But city and countryside are also backdrops for life; there the finely wrought bridge is a complement to life, a marker of community identity, and a heritage for generations, or rather, for as long as it lasts.

We cannot assume that a bridge is to be built solely for efficient transportation, defined as the process of getting people and goods as cheaply and as fast as possible from origins to destinations. In modern life, in which many feel trapped in sedentary lives, and their health suffers as a result, public places are opportunities for walking, biking, sightseeing, and recreation. The pedestrian or biker walks or bikes not necessarily to get somewhere but for exercise, health, recreation, and enjoyment. Though bridges are small percentages of the open public surface, they are nonetheless a particularly important part. They are dramatic to those who view them from a distance, and revealing to those who walk or bike on them or spend time on them. They provide special perspectives on cities, waterways, skies, and landscapes. As such they are much more than transport conduits. Like great public squares, bridges are not just objects in a place, but are in themselves public places.

In our time, we want to build cities and other human settlements that endure well in their environment over long stretches of time—let us say that is what we mean by trying to make them sustainable. A sustainable city will stay where it is and not stretch out and sprawl over the landscape. It should be planned with the expectation that it can evolve to be ever better just where it already is, providing a finer urban experience and better living conditions into the far future. And roads that connect cities, and the bridges that connect the segments of roads, should be planned on the expectation that they will continue to be there for the long run. In these respects, it is better not to treat the bridge as a short-term investment or disposable commodity.

To say so requires a confidence in the future of humanity: confidence that here, in this city, there will be people living and working for centuries to come. It is for them in the future, as well as ourselves in the present, that we should plan our infrastructures.

Those persons in the future will have to travel or send goods from place to place to trade, travel, and make a living. Roads, bridges, and rail will continue to be foundations for commerce. Bridges have to serve not merely for traffic needs projected now, but for change and development into the future. If the bridge is safe and durable, designed to appeal to the public, in a built-up settlement or along a well-travelled highway, and not harmful to the environment, it will in all likelihood prove to be of use for a long time to come. It should be built for that long-term expectation.

There is a school of thought that a future of diminishing fossil fuels will reduce the role of automobiles, and reduce the need for roads and bridges that carry traffic. But as we write, American natural gas deposits are found to be ever more abundant. And even if nonrenewable resources become scarce or more expensive, or are regulated to reduce carbon emissions, it would be foolhardy to bet that Americans would stop wanting cars, and switch en masse to transit, and hence that fewer roads will be needed. Americans will, it is a safe bet, simply be driving rechargeable cars or cars dependent on new kinds of combustion. Bridges have to be built with the expectation that they will continue to carry motor vehicles—while providing more room for bicyclists, joggers, strollers, and those just wanting to enjoy the view. The more that we want to make bridges last a long time, the more important it will be to locate them well. New technologies may help. As long as travelers' privacy can be protected, geospatial and satellite positioning systems will allow planners to be become more effective at forecasting traffic patterns, and thereby to situate bridges in the right sites.

As infrastructures go, bridges are relatively benign for the environment. With careful environmental review, sites can be selected and adjustments made to avoid harm to water, soils, birds, and habitat. Once environmental harm is minimized, the greatest single way to make the bridge more sustainable is to reduce the embodied energy within it, as by selecting materials, such as stone and reinforced concrete, that have less embodied energy. These same materials can last a long time. What is more, they prevent future embodiment of energy by averting future cycles of demolition and reconstruction. Sadly, many of the bridges now most celebrated reflect the sentiment that lightness and elegance are the most important esthetic. Many of these signature bridges will not last more than another two generations or so.

If the Romans could do it, we can: the most sustainable bridge is the one we build to last a thousand years. A long-lasting compression-based arch bridge will also more inexpensively carry additional loads—loads that would usually be considered luxuries in traditional calculations. So designed, the bridge can expand its functions to include a sitting area, garden, tree-planted boulevard, restaurant with canopies, and sheltered viewing platform. That bridge then truly becomes a well-loved place in itself.

Thanks to advances in technology, the long-lasting bridge will not have to be as heavy as the Roman ones were. New technologies of bridge health monitoring are providing real-time information on how bridges respond to varied loads, thereby improving engineers' abilities to predict what will be safe at lesser cost in the long run. The technologies will also help us better understand structures and materials conducive to long-term durability.

It will take more research to protect bridges against the greatest threats of failure—scour, flood, earthquake, and ship impact. We have to continue investing in research on the dynamics by which hazards affect structures. More work is needed on methods by which to assure that infrastructures are designed according to reliably derived and consistently calculated expectations for hazard risk and hazard intensity—across varied hazard types. The structural health monitoring systems we have just mentioned will be useful not just for research, but for real-time notification of structural problems, so bridges can be repaired or, in severe emergency, evacuated. The bridge's ability to survive a thousand years will depend on its strengthened ability to withstand extreme events.

It is not a contradiction in terms to want more long-lived bridges, yet ones delivered faster. That a bridge requires 10 to 20 years from conception to commissioning is no help to the public or to the environment. The delays and cost overruns subtract from the number and quality of infrastructure items that can be built. The remarkably lengthy delivery times deserve further careful study to help us understand where the holdups come from. But within the legal environment in which we live, there will be no easy solution. More accurate cost projections and "design-build" delivery methods may help here and there, but cannot solve the larger problem.

They cannot help us escape the fact that the United States may have taken a good idea much too far. Too many constraints, too many specialists, too many bureaucracies, and too many stakeholders stumbling over each other, causing delay upon delay—even when the proposed infrastructure is meant to reduce an environmental problem, as when the job is to replace a bridge that is causing pollution through excess congestion. Bad policy making is not a gift to the environment. In the midst of dozens of conflicting pressures and interests, funds get spent on satisfying multiple stakeholders with short-term agendas, often to settle minor environmental preferences in urban areas that are, after all, human constructions and can never be pristine. To this problem of slow delivery, we have no solution now, except to counsel further research and open public debate. We do believe that the process as it now exists diminishes the funds that could be spent for constructing finer infrastructure, including the long-lasting bridges that will create more stable, sustainable environments.

We end with our hope that citizens will call for millennial bridges. Avoiding cycles of rebuilding, each 1000-year bridge will be an optimistic commitment to permanence of place. It will be an assertion that civilization will survive and that our descendants will live here for thirty generations. It will truly be a bridge to the future.

INDEX

Note: The letter *t* following a page number denotes a table, the letter *f* a figure.

Preliminary design and environmental review (for project delivery), 144–146
Present value. *See* cost-benefit analysis.
Project delivery. *See* Delivery
PS&E. *See* Plans, specifications, and estimates

Reliability (of bridge), 60
Redundancy, 60
Riprap, 74–75

Safety Factors, 57–60
Scenarios: against terrorism, 77; for net present value analysis, 92–94; for "Square City," 116–118; in four-step model, 107, 109–110
Scoping (for project delivery), 142–144
Scour. *See* Hydraulic forces
Shaking table, 79
Shear force, 27–29
Sites (fitting bridges to), 48–50, 84, 85, 123, 129, 132, 160
Slab bridge, 39–40
Spandrel (and spandrel column), 42–43
Spans (of bridges), 9–10, 10*t*, 41*f*, 48–49; of cable-stayed bridges, 47; of suspension bridges, 45
"Square City," 107–118. *See also* Transportation modeling; Scenarios.
Stays (of cable-stayed bridges), 46–47
Steel, 37–38; embodied energy in, 133–134
Stiffness, 26–28, 37, 46
Strength (limit state), 53
Stress (and strain), 22–23, 23*f*; *defined*, 22; bending, 29–30, 30*f*, 35–36; combined: 32–33; shear, 27–29; stress maxima, 54–57; tensile, 25–27; torsional, 31–32, 35–36; strain, 22–23, 25–27; yield stress, 26, 27*f*
Strain. *See* stress.

Superstructure (and substructure), 40–41
Suspension bridge, 45–46, 45*f*
Sustainability (of bridge), 132–136; as durability, 134–135; concrete versus steel, 133–134; embodied energy, 133–135; Leadership in Energy and Environmental Design (LEED), 133; of thousand year bridge, 135

TAZ. *See* Traffic Analysis Zone
Tensile force, 25–27, 25*f*, 27*f*
Terrorism. *See* Extreme events
Thousand year bridge, 157–161; cost-benefit analysis as obstacle to, 95–96, 100; embodied energy in, 135; resisting extreme events to allow, 79; planning for, 157–161
Torsion, 31–32, 31*f*; and bending, 35–36
Traffic: analysis of, 104–106; 103–122 *passim*; delays in, 14–15; increase of, 14–15; travel demand management, 128. *See also* Transportation modeling.
Traffic Analysis Zone (TAZ), 110–116
Transportation modeling, 103–122; four-step model, 107–116; limits of, 107, 118–121; need for 106–107; "Square City," 107–118; travel cost, 86–88, 86*f*
Travel demand management, 128
Truss (and truss bridge), 43–44, 44*f*, 49, 64–65; deck-truss, 43; through-truss, 43

Viaduct, 40–41

Web (in I-beam, or box-beam), 38–39
Width. *See* dimensions

Young's modulus, 26